Benjamin Rush Field

Medical Thoughts of Shakespeare

Benjamin Rush Field

Medical Thoughts of Shakespeare

ISBN/EAN: 9783337060633

Printed in Europe, USA, Canada, Australia, Japan

Cover: Foto ©berggeist007 / pixelio.de

More available books at **www.hansebooks.com**

OF

SHAKESPEARE.

By B. RUSH FIELD, M. D.,

MEMBER OF THE SHAKESPEARE SOCIETY
OF NEW YORK.

SECOND EDITION, REVISED AND ENLARGED.

EASTON, PA.:

ANDREWS & CLIFTON, PUBLISHERS.

1885.

PREFACE TO SECOND EDITION.

If any old lady, knight, priest or physician,
Should condemn me for writing a second edition ;
If good Madam Squintum my work should abuse,
May I venture to give her a smack of my muse ?

Anstey's New Bath Guide, p. 169.

THE occasion is taken to acknowledge the kind consideration that the first edition of this little work has received. This edition appears in a thoroughly revised and much enlarged form ; to what extent, may be judged by the fact that chapters on The Physician, Surgery, Physiology, Anatomy and Pharmacy have been added, together with many allusions to the other medical subjects, making an increase of over four hundred quotations. It has been impossible to resist the temptation of adding a few medical thoughts from other authors, which will be found under their appropriate heads. The labor necessary to accomplish this has not interfered in any way with professional duties ; it being a task entirely of the leisure hours of the night.

EASTON, PENNSYLVANIA, June, 1885.

CONTENTS.

Medical Thoughts of Shakespeare.

PART I.

THE PHYSICIAN.

SHAKESPEARE'S education was not, by any means, hedged in by plots and characters; besides these, his mighty mind seems to have teemed with the knowledge of languages, medicine, law and court etiquette. It is wonderful that one brain could shine forth such a vast variety, and surprising that he has even gone into the *minutiæ* of the different avenues of learning through which he has stridden. Shakespeare paid considerable attention to medicine, and has furnished some of the finest specimens of the medical character that have ever been drawn by any writer. His Cerimon, in Pericles, is a most noble one. He speaks for himself:

> 'Tis known, I ever
> Have studied physic, through which secret art,
> By turning o'er authorities, I have
> (Together with my practice,) made familiar
> To me and to my aid, the bless'd infusions
> That dwell in vegetives, in metals, stones;
> And I can speak of the disturbances
> That nature works, and of her cures; which doth give me
> A more content in course of true delight
> Than to be thirsty after tottering honour,
> Or tie my treasure up in silken bags
> To please the fool and death.
>
> *Act III., Sc. II.*

And others speak of him:

> Hundreds call themselves
> Your creatures, who by you have been restored:
> And not your knowledge, your personal pain, but even
> Your purse, still open, hath built lord Cerimon
> Such strong renown as time shall ne'er decay.
>
> *Act III., Sc. II.*

Dowden says, "Cerimon, who is master of the secrets of nature, who is liberal in his 'learned charity,' who held it ever

> 'Virtue and cunning were endowments greater
> Than nobleness and riches,'

is like a first study of Prospero;" while Furnivall thinks that he represents to some extent the famous Stratford physician, Dr. John Hall, who married Shakespeare's eldest daughter Susanna.

What an excellent physician was Gerard de Narbon, Helena's father, who is referred to in All's Well:

> This young gentlewoman had a father, whose skill was almost as great as his honesty; had it stretched so far, would have made Nature immortal, and death should have play for lack of work. Would, for the king's sake, he were living! I think it would be the death of the king's disease. * * * * He was famous, sir, in his profession, and it was his right to be so. * * * The king * * * spoke of him admiringly and mournfully: he was skillful enough to have lived still, if knowledge could be set up against mortality.
>
> *Act I., Sc. I.*

> How long is't, count,
> Since the physician at your father's died?
> If he were living, I would try him yet;—
> * * * * * the rest have worn me out
> With several applications: nature and sickness
> Debate it at their leisure.
>
> *Act. I., Sc. II.*

> My father's skill, which was the greatest of his profession.
>
> *Act I., Sc. III.*

Another worthy physician is to be found in Cymbeline. Cornelius argues with the queen against her designs, and failing in this he completely thwarts her murderous intentions by giving her a false compound.

8

Queen. Now, master doctor, have you brought those drugs?

Cor. * * * I beseech your grace, without offence,
My conscience bids me ask,—wherefore you have
Commanded of me these most poisonous compounds,
Which are the movers of a languishing death;
But though slow, deadly?
* * * * * * * * *
   Your highness
Shall from this practice but make hard your heart:
Besides, the seeing these effects will be
Both noisome and infectious.
* * * * * * * * *
[*Aside.*] I do suspect you, madame;
But you shall do no harm.
* * * I do not like her. She doth think she has
Strange ling'ring poisons: I do know her spirit,
And will not trust one of her malice with
A drug of such damn'd nature. Those she has
Will stupify and dull the sense awhile;
* * * * * * but there is
No danger in what show of death it makes,
More than the locking up the spirits a time,
To be more fresh, reviving. She is fool'd
With a most false effect; and I the truer
So to be false with her.

          Act I., Sc. V.

The queen, sir, very oft importun'd me
To temper poisons for her; still pretending
The satisfaction of her knowledge only
In killing creatures vile, as cats and dogs,
Of no esteem: I, dreading that her purpose
Was of more danger, did compound for her
A certain stuff, which, being ta'en, would cease
The present power of life; but in short time
All offices of nature should again
Do their due function.

          Act V., Sc. V.

Macbeth supplies us with a wise member of the profession,
who, at a time when charlatans without number were promising
to cure every malady, sees clearly that Lady Macbeth's disease
is beyond his power, and so informs Macbeth.

This disease is beyond my practice:
* * * * * * infected minds

To their deaf pillows will discharge their secrets.
More needs she the divine than the physician:
* * * * * * * * *
Remove from her the means of all annoyance,
And still keep eyes upon her.

Act V., Sc. I.

King Macb. How does your patient, doctor?
Doct. Not so sick, my lord,
As she is troubled with thick-coming fancies,
That keep her from her rest.
King Macb. Cure her of that:
Canst thou not minister to a mind diseas'd;
Pluck from the memory a rooted sorrow;
Raze out the written troubles of the brain;
And, with some sweet oblivious antidote,
Cleanse the stuff'd bosom of that perilous stuff
Which weighs upon the heart?
Doct. Therein the patient
Must minister to himself.
King Macb. Throw physic to the dogs,
I'll none of it.

Macbeth, Act V., Sc. III.

In King Lear also appears a physician worthy of the name. The last scene of the fourth act shows his excellent skill in treating Lear's case. Dr. Bucknill, of England, in writing of it twenty-five years ago, says: "We confess, almost with shame, that although near two centuries and a half have passed since Shakespeare thus wrote we have very little to add to his method of treating the insane as thus pointed out."

Dr. Butts, in Henry VIII, and Dr. Caius, in Merry Wives, play rather unimportant parts. He compliments the profession by putting this speech in the mouth of a madman:

Timon to Banditti:
Trust not the physician;
His antidotes are poison, and he slays
More than you rob.

Timon of Athens, Act IV., Sc. III.

And bringing this one from the lips of an ignorant prostitute:

Nay, will you cast away your child on a fool and a physician?

Merry Wives, Act III., Sc. IV.

Reference to the physician is frequently made throughout his works.

> *Cor.* The queen is dead.
> *Cym.* Whom worse than a physician
> Would this report become. But I consider,
> By med'cine life may be prolong'd, yet death
> Will seize the doctor too.
> > *Cymbeline, Act V., Sc. V.*

> * * * * doctor-like, controlling skill.
> > *Sonnets, LXVI.*

> We * * * may not be so credulous of cure,
> When our most learned doctors leave us.
> > *All's Well, Act II., Sc. I.*

> Kill thy physician, and the fee bestow
> Upon the foul disease.
> > *King Lear, Act I., Sc. I.*

> Thou speak'st like a physician, Helicanus;
> That minister'st a potion unto me,
> That thou would'st tremble to receive thyself.
> > *Pericles, Act I., Sc. II.*

> The patient dies while the physician sleeps.
> > *Lucrece.*

> The physician
> Angry that his prescriptions are not kept,
> Hath left me.
> > *Sonnets, CXLVII.*

> Testy sick men, when their deaths be near,
> No news but health from their physicians know.
> > *Sonnets, CXL.*

> His friends, like physicians, thrice give him over.
> > *Timon of Athens, Act III., Sc. III.*

He is the wiser man, master doctor; he is a curer of souls, and you a curer of bodies.
> > *Merry Wives, Act II., Sc. III.*

> A poor physician's daughter my wife! Disdain
> Rather corrupt me ever.
> > *All's Well, Act II., Sc. III.*

> Doctors, less famous for their cures than fees.
> > *Byron—Don Juan, Canto XIV., Verse XLVIII.*

11

Like a port sculler, one physician plies
And all his art and all his skill he tries;
But two physicians, like a pair of oars,
Conduct you faster to the Stygian shores.

This is the way physicians mend or end us,
Secundum artem : but although we sneer
In health—when ill, we call them to attend us
Without the least propensity to jeer;
While that " *hiatus maxime defiendus* "
To be filled up by spade or mattock, 's near,
Instead of gliding graciously down Lethe,
We tease mild Baillie, or soft Abernethy.

Byron—Don Juan, Canto X, Verse XLII.

God and the doctor we alike adore,
But only when in danger, not before ;
The danger o'er, both are alike requited,
God is forgotten, and the doctor slighted.

The doctor says so * * * * * *
* * * * * * they sometimes
Are soothsayers and always cunning men.
Which doctor was it ?

Ben Jonson—Magnetic Lady, Act II., Sc. I.

A side thrust at the experimenters in the profession is found in Cymbeline.

I do know her spirit,
And will not trust one of her malice with
A drug of such damn'd nature. Those she has
Will stupify and dull the sense awhile ;
Which first, perchance, she'll prove on cats and dogs,
Then afterwards up higher. *Act I., Sc. V.*

I can smile, and murder whiles I smile.

Henry VI.—3d, Act III., Sc. II.

He has in several plays shown his contempt for the " prating mountebank " or " doting wizard."

They brought one Pinch, a hungry, lean-fac'd villain,
A mere anatomy, a mountebank,
A thread-bare juggler, and a fortune-teller ;
A needy, hollow-ey'd, sharp-looking wretch,
A living dead man : this pernicious slave,
Forsooth, took on him as a conjurer,
And, gazing in mine eyes, feeling my pulse,
And with no face, as 'twere, out-facing me,
Cries out I was possessed *Comedy of Errors, Act V., Sc. I.*

I say we must not
So stain our judgment, or corrupt our hope,
To prostitute our past-cure malady
To empirics; or to dissever so
Our great self and our credit, to esteem
A senseless help, when help past sense we deem.

All's Well, Act II., Sc. I.

12

PART II.

PRACTICE OF MEDICINE.

Shakespeare's maladies are many and the symptoms very well defined. Diseases of the nervous system seem to have been a favorite study, especially insanity; Lear, Timon, and Hamlet being excellent examples.

> And he * * * (a short tale to make),
> Fell into a sadness; then into a fast;
> Thence to a watch; thence into a weakness;
> Thence to a lightness; and, by this declension
> Into the madness wherein now he raves.
>
> *Hamlet, Act II., Sc. II.*

> He took me by the wrist and held me hard ;
> Then goes he to the length of all his arm ;
> And with his other hand thus o'er his brow,
> He falls to such perusal of my face,
> As he would draw it. Long stay'd he so;
> At last,—a little shaking of mine arm,
> And thrice his head thus waving up and down,
> He raised a sigh so piteous and profound,
> That it did seem to shatter all his bulk,
> And end his being: That done, he lets me go:
> And, with his head o'er his shoulder turn'd,
> He seem'd to find his way without his eyes;
> For out o' doors he went without their help,
> And, to the last, bended their light on me.
>
> *Hamlet, Act II., Sc. I.*

> Alas, how is it with you,
> That you do bend your eye on vacancy,
> And with the incorporal air do hold discourse?
> Forth at your eyes your spirits wildly peep:
> And, as the sleeping soldiers in the alarm,
> Your bedded hair, like life in excrements,
> Starts up, and stands on end.
>
> *Hamlet, Act III., Sc. IV.*

13

O, what a noble mind is here o'erthrown!
The courtier's, scholar's, soldier's, eye, tongue, sword:
The expectancy and rose of the fair state,
The glass of fashion and the mould of form,
The observed of all observers,—quite, quite down!
And I, of ladies most deject and wretched,
That suck'd the honey of his music vows,
Now see that noble and most sovereign reason,
Like sweet bells jangled, out of tune and harsh;
That unmatch'd form and feature of blown youth,
Blasted with ecstasy.

Hamlet, Act III., Sc. I.

There's something in his soul,
O'er which his melancholy sits on brood;
And I do doubt the hatch and the disclose,
Will be some danger.

Hamlet, Act III., Sc. I.

Canst thou not minister to a mind diseas'd;
Pluck from the memory a rooted sorrow;
Raze out the written troubles of the brain;
And, with some sweet oblivious antidote,
Cleanse the stuff'd bosom of that perilous stuff
Which weighs upon the heart?

Macbeth, Act V., Sc. III.

* * * * * Infected minds
To their deaf pillows will discharge their secrets.
* * * * * * * * *
Remove from her the means of all annoyance,
And still keep eyes upon her.

Macbeth, Act V., Sc. I.

Infirmity doth still neglect all office,
Whereto our health is bound; we are not ourselves,
When nature, being oppress'd, commands the mind
To suffer with the body: I'll forbear;
And am fall'n out with my more headier will,
To take the indispos'd and sickly fit
For the sound man.

King Lear, Act II., Sc. IV.

This is in thee a nature but infected;
A poor unmanly melancholy, sprung
From change of fortune.

Timon of Athens, Act IV., Sc. III.

14

The mere want of gold, and the falling-from of his friends, drove him into this melancholy.

> *Timon of Athens, Act IV., Sc. III.*

Tell him * * * * * *
* * * that his lady mourns at his disease:
Persuade him that he hath been a lunatic.

> *Taming of the Shrew, Ind., Sc. I.*

* * * Being lunatic
He rush'd into my house, and took perforce
My ring away.

> *Comedy of Errors, Act IV., Sc. III.*

These dangerous unsafe lunes.

> *Winter's Tale, Act II, Sc. II.*

With great imagination,
Proper to madmen, led his powers to death,
And, winking, leap'd into destruction.

> *Henry IV.—2d, Act. I, Sc. III.*

Oft the eye mistakes, the brain being troubled.

> *Venus and Adonis.*

To see his nobleness!
Conceiving the dishonour of his mother,
He straight declin'd, droop'd, took it deeply;
Fasten'd and fix'd the shame on 't in himself;
Threw off his spirit, his appetite, his sleep,
And downright languish'd.

> *Winter's Tale, Act II., Sc III.*

His siege is now
Against the mind, the which he pricks and wounds
With many legions of strange fantasies,
Which, in their throng and press to that last hold,
Confound themselves.

> *King John, Act V., Sc. VII.*

Shakespeare certainly had the true idea of the great value of sleep, and he also knew of its importance in the treatment of brain diseases. Sleep serves as an excellent stimulant, promoting the growth of the brain. The infant, during the first ten weeks of its life, sleeps most of the time and hence during that period its brain is overdeveloped in proportion to its size.

Our foster-nurse of nature is repose,
The which he lacks; that to provoke in him,

15

Are many simples operative, whose power
Will close the eye of anguish.
King Lear, Act IV., Sc. IV.

O sleep, gentle sleep,
Nature's soft nurse,
King Henry IV—2d, Act III., Sc. I.

Sleep, that knits up the ravell'd sleave of care,
The death of each day's life, sore labour's bath,
Balm of hurt minds, great nature's second course,
Chief nourisher of life's feast.
Macbeth, Act II., Sc. I.

Oppressed nature sleeps : —
This rest might yet have balm'd thy broken senses,
Which, if convenient will not allow,
Stand in hard cure.
King Lear, Act III., Sc. VI.

Man's rich restorative ; his balmy bath,
That supplies, lubricates and keeps in play
The various movements of that nice machine,
Which asks such frequent periods of repair.
Young's Night Thoughts.

Music was held as one of the remedies in the treatment of insanity. It plays an important part in King Lear, (IV–VII), and finds mention as a remedy in other plays.

This music mads me, let it sound no more ;
For, though it have holp madmen to their wits,
In me it seems it will make wise men mad.
Richard II., Act V., Sc. V.

Let there be no noise made, my gentle friends ;
Unless some dull and favourable hand
Will whisper music to my weary spirit.
Henry IV—2d, Act IV., Sc. IV.

Your honour's players, hearing your amendment,
Are come to play a pleasant comedy,
For so your doctors hold it very meet.
Seeing too much sadness hath congeal'd your blood,
And melancholy is the nurse of frenzy ;
Therefore, they thought it good you hear a play,
And frame your mind to mirth and merriment,
Which bars a thousand harms, and lengthens life.
Taming of the Shrew, Ind., Sc. II.

Your physicians have expressly charg'd,
In peril to incur your former malady,
That I should yet absent me from your bed.
Taming of the Shrew, Ind., Sc. II.

This closing with him fits his lunacy:
Whate'er I forge to feed his brain-sick fits,
Do you uphold and maintain in your speeches.
Titus Andronicus, Act V., Sc. II.

Dispute not with her, she is lunatic.
Richard III., Act I.. Sc III.

* * Deserves as well a dark house and a whip as madmen do.
As You Like It, Act III., Sc. II.

Why have you suffer'd me to be imprison'd,
Kept in a dark house?
Twelfth Night, Act V., Sc. I.

It is the mynde that makes good or ill,
That maketh wretch or happie, rich or poore.
Spenser—Fœrie Queene, XI-IX.

Yet they do act
Such antics and such pretty lunacies
That spite of sorrow they make you smile.
Dekker.

Grows lunatic and childish for his son.
Kyd.

When slow Disease, and all her host of pains,
Chills the warm tide which flows along the veins;
When Health, affrighted, spreads her rosy wing,
And flies with every changing gale of Spring:
Not to the aching frame alone confined,
Unyielding pangs assail the drooping mind.
Byron—Childish Recollections.

The accuracy with which Shakespeare has written of apoplexy
is justly alluded to in Bell's *Principles of Surgery*, (1815, Vol. II,
p. 557): "My readers will smile, perhaps, to see me quoting
Shakespeare among physicians and theologists; but not one of
all their tribe, populous though it be, could describe so exquis-
itely the marks of apoplexy, conspiring with the struggles for
life, and the agonies of suffocation, to deform the countenance of
the dead: so curiously does our poet present to our conception
all the signs from which it might be inferred that the good duke
Humfrey had died a violent death."

17

See, how the blood is settled in his face !
Oft have I seen a timely-parted ghost,
Of ashy semblance, meagre, pale, and bloodless,
Being all descended to the labouring heart ;
Who, in the conflict that it holds with death,
Attracts the same for aidance 'gainst the enemy ;
Which with the heart there cools, and ne'er returneth
To blush and beautify the cheek again.
But see, his face is black and full of blood ;
His eye-balls further out than when he liv'd,
Staring full ghastly like a strangled man :
His hair uprear'd, his nostrils stretch'd with struggling ;
His hands abroad display'd, as one that grasp'd
And tugg'd for life, and was by strength subdu'd.
Look on the sheets, his hair, you see, is sticking ;
His well-proportion'd beard made rough and rugged,
Like to the summer's corn by tempest lodg'd.
It can not be but he was murder'd here ;
The least of all these signs were probable.
Henry VI—2d, Act III., Sc. II.

Suddenly a grievous sickness took him,
That made him gasp, and stare, and catch the air.
Henry VI—2d, Act III., Sc. II.

Falstaff. And I hear moreover, his highness is fallen into this same whoreson
apoplexy.
Ch. Just. Well, heaven mend him ! I pray let me speak with you.
Falstaff. This apoplexy is, as I take it, a kind of lethargy, an 't to please your
lordship ; a kind of sleeping in the blood, a whoreson tingling.
Ch. Just. What tell you me of it ? Be it as it is.
Falstaff. It hath its original from much grief ; from study and perturbation
of the brain.
Henry IV—2d, Act I., Sc. II.

War. Be patient, princes ; you do know, these fits
Are with his highness very ordinary.
Stand from him, give him air ; he'll straight be well.
Clar. No, no ; he can not long hold out these pangs :
The incessant care and labour of his mind
Hath wrought the mure, that should confine it in,
So thin, that life looks through, and will break out.
     *   *   *   *   *   *   *   *   *   *
P. Humph. This apoplexy will certain be his end.
Henry IV—2d, Act IV., Sc. IV.

18

Peace is a very apoplexy, lethargy ; mulled, deaf, sleepy, insensible.
Coriolanus, Act IV., Sc. V.

Dick. Why dost thou quiver, man ?
Say. The palsy and not fear provokes me.
Cade. Nay, he nods at us, as who should say,
I'll be even with you.
Henry VI—2d, Act IV., Sc. VII.

With a palsy-fumbling on his gorget,
Shake in and out the rivet.
Troilus and Cressida, Act I., Sc. III.

How quickly should this arm of mine,
Now prisoner to the palsy, chastise thee.
Richard II, Act II., Sc. III.

Flat on the ground and still as any stone,
A very corpse, save yielding forth a breath.
Sackville.

How concisely he describes epilepsy, giving the most promi-
nent symptoms.

Casca. He fell down in the market-place, and foamed at mouth, and was
speechless.
Bru. 'Tis very like,—he has the falling sickness.
Casca. * * * * * When he came to himself again, he said, If he had
done or said anything amiss, he desired their worships to think
it was his infirmity.
Julius Cæsar, Act I., Sc. II.

Julius Cæsar was the only epileptic among his characters:
Othello is spoken of as being one, but this is merely Iago's lie to
Cassio, which is clearly shown in Othello's conversation after the
trance, it being a continuation of the former subject, which is
never the case in epilepsy.

Iago. My lord is fall'n into an epilepsy :
This is his second fit ; he had one yesterday.
Cas. Rub him about the temples.
Iago. No, forbear;
The lethargy must have his quiet course ;
If not, he foams at mouth, and by and by
Breaks out to savage madness.
Act IV., Sc. I.

A plague upon your epileptic visage !
King Lear, Act. II., Sc. II.

19

He takes some notice of the other affections classed under nervous diseases.

> Which of your hips has the most profound sciatica?
>
> *Measure for Measure, Act I., Sc. II.*

> Thou cold sciatica,
> Cripple our Senators, that their limbs may halt
> As lamely as their manners!
>
> *Timon of Athens, Act IV., Sc. I.*

> Lord, how my head aches! what a head have I!
> It beats as it would fall in twenty pieces
>
> *Romeo and Juliet, Act II., Sc. V.*

> When your head did but ache
> I knit my handkerchief about your brows.
>
> *King John, Act IV., Sc. I.*

> *Oth.* I have a pain upon my forehead here.
> *Des.* Why, that's with watching; 't will away again.
>
> *Othello, Act III., Sc. II.*

> Let our finger ache, and it indues
> Our other healthful members even to a sense
>   Of pain
>
> *Othello, Act III., Sc. IV.*

Leander, he would have lived many a fair year, though Hero had turned nun, if it had not been for a hot midsummer night; for good youth he went but forth to wash him in the Hellespont, and being taken with the cramp, was drowned.

> *As You Like It, Act IV., Sc. I.*

> The aged man that coffers-up his gold
> Is plagu'd with cramps, and gouts and painful fits.
>
> *Lucrece.*

> * * * Shorten up their sinews
> With aged cramps.
>
> *Tempest, Act IV., Sc. I.*

> To-night thou shalt have cramps,
> Side stitches that shall pen thy breath up.
>
> *Tempest, Act I., Sc. II.*

> I'll rack thee with old cramps,
> Fill all thy bones with aches.
>
> *Tempest, Act I., Sc. II.*

> Thy nerves are in their infancy again
> And have no vigour in them.
>
> *Tempest, Act I., Sc. II.*

Hysteria, in Shakespeare's time, was considered a disease common to both sexes, and was known as "*Hysterica passio*," or more popularly termed "the mother."

> O, how this mother swells up toward my heart!
> *Hysterica passio*—down, thou climbing sorrow,
> Thy element's below! Where is this daughter?
>
> *King Lear, Act II., Sc. IV.*

Percy thinks that Shakespeare read of this disease in Harsnet's "Declaration of Popish Impostures" while he was looking up material for his character of Tom of Bedlam. The following is taken from (p. 25) the work referred to : "Ma: Maynie had a spice of the *Hysterica passio* as seems from his youth, hee himself termes it the *Moother*, and saith that hee was much troubled with it in Fraunce, and that it was one of the causes that mooved him to leave his holy order whereinto he was initiated and to returne into England."

Diseases of the nervous system have not been overlooked by other writers. How excellently we have described the chief symptom of *locomotor ataxia :*

> Obliquely waddling to the mark in view.
>
> *Pope.*

And Byron well portrays vertigo.

> Her cheek turn'd ashes, ears rung, brain whirl'd round,
> As if she had received a sudden blow,
> And the heart's dew of pain sprang fast and chilly
> O'er her fair front, like morning's on a lily.
> Although she was not of the fainting sort,
> Baba thought she would faint, but there he err'd—
> It was but a convulsion, which, though short,
> Can never be described ; we all have heard,
> And some of us have felt thus "*all amort*,"
> When things beyond the common have occurr'd.
>
> *Don Juan, Canto VI., Verse CV.*

> That old vertigo in his head
> Will never leave him, till he's dead.
>
> *Swift.*

> Of all mad creatures, if the learned are right.
> It is the slaver kills and not the bite.
>
> *Pope.*

> Loss !—such a palaver,
> I'd inoculate sooner my wife with the slaver
> Of a dog when gone rabid, than listen two hours
> * * * * * *
>
> *Byron—The Blues.*

21

> The sot,
> Hath got blue devils for his morning mirrors:
> What though on Lethe's stream he seem to float,
> He can not sink his tremors or his terrors;
> The ruby glass that shakes within his hand,
> Leaves a sad sediment of Time's worst sand.
>
> *Byron—Don Juan, Canto XV., Verse IV.*

Taking up diseases of the circulatory system next we find Shakespeare displaying considerable knowledge in regard to them. The extended impulse of the heart under intense excitement is nicely shown in the Rape of Lucrece.

> His hand, that yet remains upon her breast,—
> Rude ram, to batter such an ivory wall!
> May feel her heart,—(poor citizen!) distress'd.
> Wounding itself to death, rise up and fall,
> Beating her bulk, that his hand shakes withal.

Again,

> I fear'd thy fortune, and my joints did tremble.
> *   *   *   *   *   *   *
> My boding heart pants, beats, and takes no rest.
> But, like an earthquake, shakes thee on my breast.
>
> *Venus and Adonis.*

> I have *tremor cordis* on me,—my heart dances.
>
> *Winter's Tale, Act I., Sc. II.*

> Whose horrid image doth unfix my hair,
> And make my seated heart knock at my ribs,
> Against the use of nature?
>
> *Macbeth, Act I., Sc. III.*

Death from "broken heart," caused by excessive grief, finds mention in several plays.

> Woe the while!
> O, cut my lace; lest my heart, cracking it,
> Break too!
>
> *Winter's Tale, Act III., Sc. II.*

> The grief that does not speak,
> Whispers the o'er-fraught heart, and bids it break.
>
> *Macbeth, Act IV., Sc. III.*

> Shall split thy very heart with sorrow.
>
> *Richard III., Act I., Sc. III.*

22

Dyer in his "Folk-Lore of Shakespeare" quotes the following from Mr. Timb's "Mysteries of Life, Death, and Futurity," (1861, p. 149.) "This affection (broken-heart) was, it is believed, first described by Harvey; but since his day several cases have been observed. Morgagni has recorded a few examples: among them, that of George II., who died in 1760; and, what is very curious, he fell a victim to the same malady. Dr. Elliotson, in his Lumleyan Lectures on Diseases of the Heart, in 1839, stated that he had only seen one instance; but in the 'Cyclopædia of Practical Medicine' Dr. Townsend gives a table of twenty-five cases, collected from various authors."

A very good case of syncope is presented in Pericles. "The cases of apparent death, in which it is believed that premature interment sometimes takes place, are of this kind. Instances have occurred in which the pulse, respiration and consciousness have been absent for several days, and yet the patient has ultimately recovered. The system is in a sort of hybernation, in which vitality remains, though the vital functions are suspended. It is probable that, in such cases, a very careful auscultation might detect a slight sound in the heart." (Dr. George B. Wood's Practice. 1858. Vol. II., p. 211.)

> Make a fire within;
> Fetch hither all my boxes in my closet.
> Death may usurp on nature many hours,
> And yet the fire of life kindle again
> The o'erpress'd spirits. I have heard
> Of an Egyptian that had nine hours lien dead,
> Who was by good appliance recovered.
> * * * * * the fire and cloths—
> The rough and woeful music that we have,
> Cause it to sound, 'beseech you.
> The viol once more; * * *
> * * * I pray you, give her air;
> This queen will live; nature awakes; a warmth
> Breathes out of her: She hath not been entranc'd
> About five hours. See how she 'gins to blow
> Into life's flower again!
> * * * * * * * * . *
> Hush, my gentle neighbors!
> Lend me your hands; to the next chamber bear her.

Get linen; now this matter must be looked to,
For her relapse is mortal. Come, come,
And Æsculapius guide us!

Act III., Sc. II.

Take thou this phial, being then in bed,
And this distilled liquor drink thou off':
When, presently, through all thy veins shall run
A cold and drowsy humour, for no pulse
Shall keep his native progress, but surcease,
No warmth, no breath, shall testify thou liv'st;
The roses in thy lips and cheeks shall fade
To paly ashes; thy eyes' windows fall,
Like death, when he shuts up the day of life;
Each part, depriv'd of supple government,
Shall, stiff, and stark, and cold, appear like death:
And in this borrow'd likeness of shrunk death
Thou shalt continue two and forty hours,
And then awake as from a pleasant sleep.

Romeo and Juliet, Act IV., Sc. I.

Why does my blood thus muster to my heart,
Making both it unable for itself,
And disposessing all my other parts
Of necessary fitness?
So play the foolish throngs with one that swoons;
Come all to help him, and so stop the air
By which he should revive.

Measure for Measure, Act II., Sc. IV.

Many will swoon when they do look on blood.

As You Like It, Act. IV., Sc. III.

No damsel faints when rather closely press'd,
But more caressing seems when most caress'd;
Superfluous hartshorn, and reviving salts,
Both banish'd by the sovereign cordial " waltz."

Byron—The Waltz.

Some attention has been paid to chlorosis:

Out, you green-sickness carrion! Out, you baggage,
You tallow-face!

Romeo and Juliet, Act III, Sc. V.

Pand. The pox upon her green sickness for me.
Bawd. Faith, there's no way to be rid on 't, but by the way to the pox.

Pericles, Act IV., Sc. VI.

24

There's never any of these demure boys come to any proof; for thin drink
doth so overcool their blood, and making many fish-meals, that they fall into
a kind of male green sickness; they are generally fools and cowards.

Henry IV—2d, Act IV., Sc. III.

Lepidus,
Since Pompey's feast, as Menas says, is troubled
With the green sickness.

Antony and Cleopatra, Act III., Sc. II.

Ben Jonson in writing of this disease has happily and properly
recommended marriage as an important step toward recovery.

He would keep you * * * not alone without a husband,
But with a sickness; ay, and the green sickness,
The maiden's malady; which is a sickness,—
A kind of a disease, * * * * *
And like the fish our mariners call *remora*.
* * * * * * * *
I say remora,
For it will stay a ship that's under sail;
And stays are long and tedious things to maids!
And maids are young ships that would be sailing
When they be rigg'd. * * * * *
The stay is dangerous.
* * * * * * * *
I can assure you from the doctor's mouth,
She has a dropsy, and must change the air
Before she can recover.
* * * * * * * *
Give her vent.
If she do swell. A gimblet must be had:
It is a tympanites she is troubled with.
There are three kinds: the first is amsarea.
Under the flesh a tumor; that's not hers.
The second is ascites, or aquosus,
A watery humour; that is not hers neither;
But tympanites, which we call the drum,
A wind-bombs in her belly, must be unbraced,
And with a faucet or a peg, let out,
And she'll do well: get her a husband.

Magnetic Lady, Act II., Sc. I.

My nose fell a-bleeding on Black-Monday last.

Merchant of Venice, Act II., Sc. V.

Diseases of the respiratory system were quite overlooked by
Shakespeare.

Consumption catch thee!

Timon of Athens, Act IV., Sc. III.

There's hell, there's darkness, there is the sulphurous pit, burning, scalding, stench, consumption!

King Lear, Act IV., Sc. VI.

Thy food is such
As has been belch'd on by infected lungs.

Pericles, Act IV., Sc. VI.

But I'm relapsing into metaphysics,
That labyrinth, whose clue is of the same
Construction as your cures for hectic phthisics,
Those bright moths fluttering round a dying flame.

Byron—Don Juan, Canto XII., Verse LXXII.

Love is riotous, but marriage should have quiet,
And, being consumptive, live on a milk diet.

Byron—Don Juan, Canto XV., Verse XLI.

For goodness, growing to a plurisy,
Dies in his own too-much.

Hamlet, Act IV., Sc. VII.

A whoreson cold, sir; a cough, sir; which I caught with ringing in the king's affairs, upon his coronation day.

Henry IV—2d, Act III., Sc. II.

'Tis dangerous to take a cold.

Henry IV., Act II., Sc. III.

The tailor cries, and falls into a cough.

Midsummer Night's Dream, Act II., Sc. I.

Coughs will come when sighs depart.

Byron—Don Juan, Canto X., Verse VIII.

Who, * * * but would much rather
Sigh like his son, than cough like his grandfather?

Byron—Don Juan, Canto X., Verse VI.

He has not forgotten the diseases affecting the digestive organs.

An old superstition regarding toothache was that it was caused by a small worm, formed like an eel, which bored a hole into the tooth, and various methods were employed to remove it. Dyer. notes the fact that John of Gatisden, one of the oldest medical authorities, attributed decay of the teeth to this cause.

Don Pedro. What! sigh for the toothache?
Leon. Where is but a humour or a worm?

Much Ado, Act III., Sc. II.

He that sleeps feels not the toothache.

Cymbeline, Act V., Sc. IV.

Being troubled with a raging tooth,
　I could not sleep.
Othello, Act III., Sc. III.

There was never yet philosopher,
That could endure the toothache patiently.
Much Ado, Act V., Sc. I.

She shall be buried with her face upwards;
Yet this is no charm for the toothache.
Much Ado, Act III., Sc. II.

Bene. I have the toothache.
D. Pedro. Draw it.
Much Ado, Act III., Sc. II.

Things sweet to taste prove in digestion sour.
Richard II., Act I, Sc. III.

A surfeit of the sweetest things
The deepest loathing to the stomach brings.
Midsummer Night's Dream, Act II., Sc. II.

Like a sickness, did I loath this food:
But, as in health, come to my natural taste,
Now do I wish it, love it, long for it.　*　*
Midsummer Night's Dream, Act IV, Sc. I.

　She gallops night by night.　*　*
*　*　*　*　*　*　*　*
O'er ladies lips, who straight on kisses dream;
Which oft the angry Mab with blisters plagues,
Because their breaths with sweetmeats tainted are.
Romeo and Juliet, Act I., Sc. IV.

Fat paunches have lean pates, and dainty bits
Make rich the ribs, but bankrupt quite the wits.
Love's Labour's Lost, Act I., Sc. I.

Say, can you fast? Your stomachs are too young;
And abstinence engenders maladies.
Love's Labour's Lost, Act IV., Sc. III.

Unquiet meals make ill digestions.
Comedy of Errors, Act V., Sc. I.

A sick man's appetite, who desires most that
Which would increase his evil.
Coriolanus, Act I., Sc. I.

Do not turn me about; my stomach is not constant.
Tempest, Act II., Sc. II.

27

For, ever and anon comes indigestion.
Byron—Don Juan, Canto XI., Verse III.

When a roast and a ragout,
And fish and soup, by some side-dishes back'd,
Can give us either pain or pleasure, who
Would pique himself on intellects, whose use
Depends so much upon the gastric juice?
Byron—Don Juan, Canto V., Verse XXXII.

He ate and he was well supplied; and she
Who watch'd him like a mother, would have fed
Him past all bounds, because she smiled to see,
Such appetite in one she had deem'd dead:
But Zoe, being older than Haidee,
Knew (by tradition, for she ne'er had read),
That famish'd people must be slowly nursed,
And fed by spoonfuls, else they always burst.
Byron—Don Juan, Canto II., Verse CLVIII.

Why look you pale?
Seasick, I think, coming from Muscovy.
Love's Labour's Lost, Act V., Sc. II.

The shepherd's daughter * * * who began to be much seasick.
Winter's Tale, Act V., Sc. II.

—— the impatient wind blew half a gale:
High dash'd the spray, the bows dipp'd in the sea,
And seasick passengers turn'd somewhat pale.
Byron—Don Juan, Canto X., Verse LXIV.

Now we've reached her, lo! the captain,
Gallant Kidd, commands the crew;
Passengers their berths are clapt in,
Some to grumble, some to spew.
* * * * * *
"Help!"—"a couplet?"—"no, a cup
Of warm water."
"What's the matter?"
"Zounds! my liver 's coming up;
I shall not survive the racket
Of this brutal Lisbon Packet."
Byron — Poems.

Love 's a capricious power; I've known it hold
Out through a fever caused by its own heat,
But be much puzzled by a cough or cold,
And find a quinsy very hard to treat;
Against all noble maladies he 's bold,
But vulgar illnesses don't like to meet,
Nor that a sneeze should interrupt his sigh,
Nor inflammations redden his blind eye.
But worst of all it's nausea, or a pain
About the lower regions of the bowels:
Love who heroically breathes a vein,
Shrinks from the application of hot towels,

28

And purgatives are dangerous to his reign,
Seasickness death.
                    Byron—Don Juan, Canto II., Verse XXII.

Like wind compress'd and pent within a bladder,
Or like a human colic which is sadder.
                    Byron—Vision of Judgment.

When will your constipation have done, good madame?
                    Cartwright.

Diseases of the secretory system have not escaped his eagle eye.

A fat old man * * * that swoln parcel of dropsies.
                    Henry IV., Act II., Sc. IV.

    The dropsy drown this fool!
                    Tempest, Act IV., Sc. I.

    It is a dropsied honour.
                    All's Well, Act II., Sc. III.

Fal. You make fat rascals, mistress Doll.
Doll. I make them! gluttony and disease make them.
                    Henry IV—2d, Act II., Sc. IV.

Leprosy was sometimes called measles, from the French of leper, *meseau* or *mesel.* This is the sense in which Shakespeare uses the word measles—an entirely different one from that now in vogue. The word "hoar," occurring in several of the quotations, refers to the white spots so characteristic of the disease.

    As for my country I have shed my blood,
    Not fearing outward force, so shall my lungs
    Coin' words till their decay against those measles,
    Which we disdain should tetter us, yet sought
    The very way to catch them.
                    Coriolanus, Act III., Sc. I.

    Gold! * * * * * *
    This yellow slave will make the hoar leprosy ador'd.
                    Timon of Athens, Act IV., Sc. III.

        Hoar the flamen,
    That scolds against the quality of flesh,
    And not believes himself.
                    Timon of Athens, Act IV., Sc. III.

        Itches, blains,
    Sow all the Athenian bosoms, and their crop
        Be general leprosy!
                    Timon of Athens, Act IV., Sc. I.

29

Diseased nature oftimes breaks forth
In strange eruptions.

Henry IV., Act III., Sc. I.

For thine own bowels, which do call thee sire,
The mere effusion of thy proper loins,
Do curse the gout, *serpigo*, and the rheum,
For ending thee no sooner.

Measure for Measure, Act III., Sc. I.

Now the dry scripgo on the subject!

Troilus and Cressida, Act II., Sc. III.

A tailor might scratch her where 'er she did itch.

Tempest, Act II., Sc. II.

In the midland counties of England a pimple was frequently
called " a quat."

I have rubb'd this young quat almost to a sense,
And he grows angry.

Othello. Act V., Sc. I.

Rubbing the poor itch,
* * * Make yourselves scabs

Coriolanus, Act I., Sc. I.

I would thou didst itch from head to foot, and I had the scratching of thee;
I would make thee the loathsomest scab in Greece.

Troilus and Cressida, Act II., Sc. I.

My elbow itched; I thought there would a scab follow.

Much Ado, Act III., Sc. III.

Scratching her legs that one shall swear she bleeds.

Taming of the Shrew, Ind., Sc. II.

Full of unpleasing blots and sightless stains.

King John, Act III., Sc. I.

Dro. S. She sweats—a man may go over shoes in the grime of it.
Ant. S. That's a fault that water will mend.
Dro. S. No, sir, 'tis in grain.

Comedy of Errors, Act III., Sc. II.

I had rather heat my liver with drinking.

Antony and Cleopatra, Act I., Sc. II.

Let my liver rather heat with wine,
Than my heart cool with mortifying groans.

Merchant of Venice, Act I., Sc. I.

Were my wife's liver
Infected as her life, she would not live
The running of one glass.
Winter's Tale, Act I., Sc. II.

What grief hath set the jaundice on your cheeks?
Troilus and Cressida, Act I., Sc. III.

All seems infected that the infected spy,
And all seems yellow to the jaundiced eye.

The liver is the lazaret of bile,
But very rarely executes its function,
For the first passion stays there such a while
That all the rest creep in and form a junction,
Like knots of vipers on a dunghill's soil,
Rage, fear, hate, jealousy, revenge, compunction,
So that all mischiefs spring up from this entrail,
Like earthquakes from the hidden fire call'd "central."
Byron—Don Juan, Canto III., Verse CCXV.

The examination of the urine as an aid to diagnosis has been resorted to for many centuries, but the processes of to-day are, of course, vastly different from and hardly to be compared with those of earlier times, when blind ignorance caused urine-examining, or "*water-casting*," to be a mere mockery. The practice, says Dr. Bucknill, arose " like the barber surgery, from the ecclesiastical interdics upon the medical vocations of the clergy. Priests and monks, being unable to visit their former patients, are said first to have resorted to the expedient of divining the malady, and directing the treatment upon simple inspection of the urine." The College of Physicians, in an old statute, denounced it as belonging only to charlatans, and members were not allowed to give advice on inspection only. Shakespeare has frequently referred to it, as have also many others of the old writers, who condemn strongly what was then a shallow deception, but what has now become, by the light of knowledge, one of the most important diagnostic aids to many diseases.

Host. Thou art a Castilian, king urinal!
* * * Pardon, a word, monsieur, mock-water.
Dr. Caius. Mock-vater! vat is dat ?
Merry Wives, Act II., Sc. III.

If thou could'st, doctor, cast
The water of my land, find her disease,

31

And purge it to a sound and pristine health,
I would applaud thee to the very echo.

Macbeth, Act V., Sc. III.

Carry his water to the wise woman.

Twelfth Night, Act III., Sc. IV.

Falstaff. What says the doctor to my water?
Page. He said, sir, the water itself was a good healthy water; but, for the party that owed it, he might have more diseases than he knew for.

Henry IV—2d, Act I., Sc. II.

Others, when the bagpipe sings i' the nose
Cannot contain their urine: for affection,
Master of passion, sways it to the mood
Of what it likes or loathes.

Merchant of Venice, Act IV., Sc. I.

Macd. What three things does drink especially provoke?
Port. Marry, sir, nose-painting, sleep, and urine.

Macbeth, Act II., Sc. II.

When he makes water, his urine is congealed ice.

Measure for Measure, Act III., Sc. II.

Fevers and other general diseases are often referred to and very many excellent allusions have been made to them.

He is so shaked of a burning quotidian tertian, that it is most lamentable to behold.

Henry V., Act II., Sc. I.

If all the wine in my bottle will recover him, I will help his ague.

Tempest, Act II., Sc. II.

A lunatic lean-witted fool,
Presuming on an ague's privilege,
Dar'st with thy frozen admonition
Make pale our cheek; chasing the royal blood,
With fury, from his native residence.

Richard II., Act II., Sc. I.

But now will canker sorrow eat my bud,
And chase the native beauty from his cheek,
And he will look as hollow as a ghost,
As dim and meagre as an ague's fit,
And so he'll die.

King John, Act III., Sc. IV.

Here let them lie till famine and the ague eat them up.

Macbeth, Act V., Sc. V.

32

An untimely ague
Stay'd me a prisoner in my chamber.
Henry VIII., Act I., Sc. I.

My wind * * * would blow me to an ague.
Merchant of Venice, Act I., Sc. I.

He had a fever when he was in Spain,
And, when the fit was on him, I did mark
How he did shake; 'tis true, this god did shake:
His coward lips did from their colour fly;
And that same eye whose bend did awe the world
Did lose his lustre: I did hear him groan:
Ay, and that tongue of his, that bade the Romans
Mark him, and write his speeches in their books,
Alas! it cried, *Give me some drink, Titinius,*
As a sick girl.
Julius Cæsar, Act I., Sc. II.

Home without boots, and in foul weather too!
How 'scapes he agues?
Henry IV., Act III., Sc. I.

Danger, like an ague, subtly taints
Even then when we sit idly in the sun.
Troilus and Cressida, Act III., Sc. III.

All the infections that the sun sucks up
From bogs, fens, flats, on Prosper fall, and make him
By inch-meal a disease!
Tempest, Act II., Sc. II.

It is not for your health thus to commit
Your weak condition to the raw cold morning.
Julius Cæsar, Act II., Sc. I.

I asked the doctors after his disease—
He died of the slow fever called the tertian,
And left his widow to her own aversion.
Byron—Don Juan, Canto I., Verse XXXIV.

His feelings had not those strange fits, like tertians
Of common likings, which make some deplore
What they should laugh at—the mere ague still
Of men's regards, the fever or the chill.
Byron—Don Juan, Canto XIII., Verse XVII.

Plague has been alluded to frequently, but generally only
the symptoms of carbuncles and the petechiæ are mentioned.
As the latter only occur in very bad cases, they were called
"God's tokens," and their appearance denoted a fatal termina-

33

tion of the disease. Hence the home of the patient was closed and " Lord have mercy on us " placed upon the door.

> Write *Lord have mercy on us* on those three;
> They are infected, in their hearts it lies;
> They have the plague and caught it of your eyes.
>> *Love's Labour's Lost, Act V., Sc. II.*

> He is so plaguy-proud, that the death tokens of it cry—
> *No recovery.*
>> *Troilus and Cressida, Act II., Sc. III.*

Enobarbus. How appears the fight?

Scarus.   On our side like the token'd pestilence,
>   Where death is sure
>> *Antony and Cleopatra, Act III., Sc. X.*

> Now the red pestilence strike all trades in Rome,
> And occupations perish!
>> *Coriolanus, Act IV., Sc. I.*

> The searchers of the town,
> Suspecting that we both were in a house
> Where the infectious pestilence did reign,
> Sealed up the doors and would not let us forth.
>> *Romeo and Juliet, Act V., Sc. II.*

> Thou art a boil,
> A plague sore, an embossed carbuncle,
> In my corrupted blood.
>> *King Lear, Act II., Sc. IV.*

> Boils and plagues
> Plaster you o'er; that you may be abhorr'd
> Further than seen, and one infect another
> Against the wind a mile!
>> *Coriolanus, Act I., Sc. IV.*

> Men take diseases, one of another:
> Therefore, let men take heed of their company.
>> *Henry IV—2d, Act V., Sc. I.*

> Being sick * * * * * *
> And as the wretch, whose fever-weaken'd joints,
> Like strengthless hinges, buckle under life.
>> *Henry IV—2d, Act I., Sc. I.*

> We are all diseas'd ; and
> * * * * * * * * *
> Have brought ourselves into a burning fever,
> And we must bleed for it.
>> *Henry IV—2d, Act IV., Sc. I.*

This fever, that hath troubled me so long,
Lies heavy on me. * * * *
This tyrant fever burns me up,
And will not let me welcome this good news.

King John, Act V., Sc. III.

What's a fever but a fit of madness?

Comedy of Errors, Act V., Sc. I.

At this instant he is sick, my lord,
Of a strange fever.

Measure for Measure, Act V., Sc. I.

My heart beats thicker than a feverous pulse.

Troilus and Cressida, Act III., Sc. II.

Sickness is catching.

Midsummer Night's Dream, Act I., Sc. I.

Thus saith the preacher : "Nought beneath the sun,
Is new," yet still from change to change we run :
What varied wonders tempt us as they pass!
The cow-pox, tractors, galvanism, and gas,
In turns appear, to make the vulgar stare,
Till the swoln bubble bursts—and all is air!

Byron—Eng. Bards and Scotch Reviewers.

Vaccination certainly has been
A kind antithesis to Congreve's rockets,
With which the Doctor paid off an old pox,
By borrowing a new one from an ox

Byron—Don Juan, Canto I., Verse CXXIX.

I don't know how it was, but he grew sick :
The empress was alarm'd, and her physician
(The same who physick'd Peter), found the tick
Of his fierce pulse betoken a condition
Which augur'd of the dead, however *quick*
Itself, and show'd a feverish disposition ;
At which the whole court was extremely troubled,
The sovereign shock'd, and all his medicines doubled.
Low were the whispers, manifold the rumours :
Some said he had been poison'd by Potemkin ;
Others talked learnedly of certain tumours,
Exhaustion, or disorders of the same kin ;
Some said 'twas a concoction of the humours,
With which the blood too readily will claim kin ;
Others again were ready to maintain,
"'Twas only the fatigue of last campaign."
But here is one prescription out of many (
"Sodæ-sulphat. 3. VI. 3. S. mannæ optim.
Aq. fervent. F. 3. iss. 3. ij tinct. sennæ
Haustus," (and here the surgeon came and cupp'd him),
R. Pulv. com. gr iii. Ipecacuanhæ,"
(With more besides, if Juan had not stopp'd 'em).

" Bolus potassæ sulphuret. sumendus,
Et haustus ter in die capiendus."
This is the way physicians mend or end us,
Secundum artem. * * * * *
Byron—Don Juan, Canto X., Verse XXXIX.

Rheumatic diseases do abound :
And through this distemperature, we see
The seasons alter.
Midsummer Night's Dream, Act II., Sc. I.

This raw rheumatic day.
Merry Wives, Act III., Sc. I.

Is Brutus sick,—and is it physical
To walk unbraced, and suck up humours
Of the dank morning?   What, is Brutus sick,
And will he steal out of his wholesome bed,
To dare the vile contagion of the night,
And tempt the rheuma and unpurged air
To add unto his sickness?
Julius Cæsar, Act II., Sc. I.

Is this the poultice for my aching bones ?
Romeo and Juliet, Act II., Sc. V.

A coming shower your shooting corns presage,
Old aches will throb, your hollow tooth will rage.
Swift.

Yet am I better
Than one that's sick o' the gout, since he had rather
Groan so in perpetuity, than be cur'd
By the sure physician, death.
Cymbeline, Act V., Sc. IV.

A rich man that hath not the gout.
As You Like It, Act III., Sc. II.

His grace was rather pained
With some slight, light, hereditary twinges
Of gout, which rusts aristocratic hinges.
Byron—Don Juan, Canto, XVI., Verse XXXIV.

It is a hard, although a common case,
To find our children running restive—they
In whom our brightest days we would retrace,
Our little selves reform'd in finer clay ;
Just as old age is creeping on apace,
And clouds come o'er the sunset of our day,
They kindly leave us, though not quite alone,
But in good company—the gout and stone.
Byron—Don Juan, Canto III., Verse LIX.

Life's thin thread 's spun out
Between the gaping heir and gnawing gout.
Byron—Don Juan, Canto XIII., Verse XL.

Dear honest Ned is in the gout,
Lies racked with pain, and you without :
How patiently you hear him groan !
How glad the case is not your own!
* * * * * * *
Yet should some neighbor feel a pain
Just in the parts where I complain,
How many a message would he send !
What hearty prayers that I should mend !
Inquire what regimen I kept?
What gave me ease, and how I slept ?
And more lament when I was dead,
Than all my snivellers round my bed.

Swift—" Death of Dr. Swift."

Diseases of the absorbent system are well represented by scrofula, or " King's evil," as it was known in Shakespeare's time. This disease, so called on account of the supposed power of cure being invested in the handling and prayers of the king, was first so treated by Edward the Confessor, in 1058, and by all the succeeding rulers until William III., who refused. Queen Anne resumed the practice, but King George I. put an end to it. During the twenty years following 1662 upwards of 100,000 persons were touched for the malady.

Malcolm. Comes the king forth I pray you?
Doctor.   Ay, sir; there are a crew of wretched souls
    That stay his cure; their malady convinces
    The great assay of art; but, at his touch,
    Such sanctity hath heaven given his hand,
    They presently amend.
Malcolm.   I thank you, doctor.
Macduff. What's the disease he means?
Malcolm.   'Tis call'd the evil
    A most miraculous work in this good king :
    Which often, since my here-remain in England,
    I have seen him do. How he solicits heaven,
    Himself best knows: but strangely-visited people,
    All swoln and ulcerous, pitiful to the eye,
    The mere despair of surgery, he cures ;
    Hanging a golden stamp about their necks,
    Put on with holy prayers ; and 'tis spoken,
    To the succeeding royalty he leaves
    The healing benediction.

Macbeth, Act IV., Sc. III.

On the action of medicines he has given us abundant cause to think he was much better informed than the average man of his time.

> *Cleo.* Give me to drink mandragora
> *Char.* Why, madame?
> *Cleo.* That I might sleep out this great gap of time,
> My Antony is away.
> > *Antony and Cleopatra, Act I., Sc. V.*

> Not poppy, nor mandragora,
> Nor all the drowsy syrups of the world,
> Shall ever med'cine thee to that sweet sleep
> Which thou ow'dst yesterday.
> > *Othello, Act III., Sc. III.*

> Cupid's cup
> With the first draught intoxicates apace—
> A quintessential laudanum or "black drop"
> Which makes one drunk at once, without the base
> Expedient of full bumpers.
> > *Byron—Don Juan, Canto IX,. Verse LXVII.*

> ——like an opiate which brings troubled rest,
> Or none,
> > *Byron—Don Juan, Canto XVI., Verse X*

> The drug he gave me, which, he said, was precious
> And cordial to me, have I not found it
> Murderous to the senses?
> > *Cymbeline, Act IV., Sc. II.*

> Have we eaten of the insane root,
> That takes the reason prisoner?
> > *Macbeth, Act I., Sc. III.*

Commentators think that Shakespeare found the name of this root in Bateman's Commentary on Bartholeme *de Propriet Rerum :* "Henbane (Hyoscyamus) is called *Insana,* mad, for the use thereof is perillous; for if it be eate or drunke, it breedeth madnesse, or slow lykenesse of sleepe. Therefore this hearb is called commonly Mirilidium, for it taketh away wit and reason."
> > *Lib. XVII., Ch. 87.*

> Thy uncle stole,
> With juice of cursed hebenon in a vial,
> And in the porches of mine ears did pour
> The leperous distilment; whose effect
> Holds such an enmity with blood of man,
> That, swift as quicksilver, it courses through
> The natural gates and alleys of the body;

And with a sudden rigour, it doth posset
And curd, like sour droppings into milk,
The thin and wholesome blood: so did it mine,
And a most instant tetter bark'd about,
Most lazar-like, with vile and loathsome crust,
All my smooth body.

Hamlet, Act I., Sc. V.

It would indeed be interesting to know the source of Shakespeare's knowledge on the physiological action of this alkaloid of tobacco. Most true it is that he has selected an excellent drug for his purpose in taking up the crude oil—Nicotia nin (hebenon). Birds will fall dead as they approach it; one drop is sufficient to kill a dog; and man dies in from two to five minutes after taking a poisonous dose: but the drug produces death by the *failure of respiration*, not by its direct action on the blood. "In nicotia-poisoning the blood is, however, not perceptibly affected. The amount of the alkaloid necessary to take life is exceedingly small, and although death by asphyxia causes the vital fluid to be everywhere dark, yet the microscope reveals only normal corpuscles. Moreover, Krocker has found that the dark blood rapidly assumes an arterial hue when shaken in the air, and that its spectrum is normal." (H. C. Wood's Toxicology, 1882, p. 370.) It is thought by many that Shakespeare did not intend "hebenon" to mean the alkaloid of tobacco, and very plausible arguments have been brought forward to show that he meant hebon or the juice of the yew. Dyer, in his chapter on plants, gives the following extract of a paper read by Rev. W. A. Harrison before the New Shakespeare Society in 1882: "It has been suggested that the poison intended by the Ghost in 'Hamlet,' (I-V.), when he speaks of the 'juice of cursed hebenon,' is that of the yew, and is the same as Marlowe's 'juice of hebon.' (Jew of Malta, III-IV.) The yew is called hebon by Spenser and by other writers of Shakespeare's age; and in its various forms of eben, eiben, hiben, etc., this tree is so named in no less than five different European languages. From medical authorities, both of ancient and modern times, it would seem that the juice of the yew is a rapidly fatal poison; next, that the symptoms attending upon yew-poisoning correspond, in a very remarkable manner, with those which follow the bites of poisonous snakes; and,

39

lastly, that no other poison but the yew produces the "lazar-like ulcerations on the body, upon which Shakespeare, in this passage, lays so much stress." From these arguments there seems to be every reason for believing that Shakespeare did mean the juice of the yew, and it is to be hoped that the continual harping on this subject, as an evidence of his medical ignorance, will soon cease.

> Recovered again with aquavitæ, or some other hot infusion.
> *Winter's Tale, Act IV., Sc. III.*

> I must needs wake you: * * * *
> Alas! my lady's dead! * * * *
> * * * * * Some aquavitæ, ho!
> *Romeo and Juliet, Act IV., Sc. V.*

The second property of your excellent sherris is—the warming of the blood; which, before cold and settled, left the liver white and pale, * * * but the sherris warms it, and makes it course from the inwards to the parts extreme.
> *Henry IV—2d, Act IV., Sc. III.*

The rapidity with which aconite, in poisonous doses, acts, is forcibly shown in the comparison of it with gunpowder.

> A hoop of gold to bind thy brothers in,
> That the united vessel of their blood,
> Mingled with venom of suggestion,
> (As, force perforce, the age will pour it in,)
> Shall never leak, though it do work as strong
> As aconitum, or rash gunpowder.
> *Henry IV—2d, Act IV., Sc. IV.*

> Let me have
> A dram of poison; such soon-speeding gear
> As will disperse itself through all the veins,
> That the life-weary taker may fall dead;
> And that the trunk may be discharg'd of breath
> As violently, as hasty powder fir'd
> Doth hurry from the fatal cannon's womb.
> *Romeo and Juliet, Act V., Sc. I.*

The curative properties of balm or balsam have been known and valued for ages past.

> But, saying thus, instead of oil and balm,
> Thou lay'st in every gash that love hath given me
> The knife that made it.
> *Troilus and Cressida, Act I., Sc. I.*

Is this the balsam that the usuring senate
Pours into captain's wounds?   Banishment!
                        Timon of Athens, Act III., Sc. V.

My pity hath been balm to heal their wounds.
                        Henry VI.—3d, Act IV., Sc. III.

A solution of gold was supposed to possess great medical
power; even the actual contact of the pure metal, according to
their belief, kept the wearer ever in good health. Dyer quotes
from John Wight's translation of the "Secrets of Alexis," in
which is given a receipt " to dissolve and reducte golde into a
potable licour which conserveth the youth and healthe of a man,
and will heale every disease that is thought incurable in the
space of seven daies at the furthest." The term "grand liquor,"
as it appears in Shakespeare, refers to this solution.

Coming to look on you, thinking you dead,
(And dead almost, my liege, to think you were,)
I spake unto the crown, as having sense,
And thus upbraided it: *The care on thee depending,*
Hath fed upon the body of my father ;
Therefore, thou, best of gold, art worst of gold ;
Other, less fine in carat, is more pretious,
Preserving life in med'cine potable.
                        Henry IV—2d, Act IV., Sc. IV.

        Plutus himself,
That knows the tinct and multiplying medicine,
Hath not in nature's mystery more science
Than I have in this ring.
                        All's Well, Act V., Sc. III.

Find this grand liquor that hath gilded 'em.
                        Tempest, Act V., Sc. I.

We sicken to shun sickness when we purge.
                        Sonnets, CXVIII.

What rhubarb, senna, or what purgative drug,
Would scour these English hence?
                        Macbeth, Act V., Sc. III.

Let's purge this choler without letting blood:
This we prescribe, though no physician ;
*   *   *   *   *   *   *   *   *   *
Our doctors say, this is no month to bleed.
                        Richard II., Act I., Sc. I.

41

That gentle physic, given in time, had cur'd me;
But now I am past all *   *   *
> *Henry VIII., Act IV., Sc. II.*

'Tis time to give 'em physic, their diseases
Are grown so catching.
> *Henry VIII., Act I., Sc. III.*

He brings his physic
After his patient's death.
> *Henry VIII., Act III., Sc. II.*

I will not cast away my physic, but on those that are sick.
> *As You Like It, Act III., Sc. II.*

To jump a body with a dangerous physic
That's sure of death without it.
> *Coriolanus, Act III., Sc. I.*

Doctors give physic by way of prevention.
> *Swift.*

The ignorant and superstitious were of the opinion that poisons could be prepared so that the effect could be produced at certain periods after their ingestion. They were also in error in the thought that poisons caused great swelling of the body.

She did confess she had
For you a mortal mineral; which, being took,
Should by the minute feed on life, and, lingering,
By inches waste you.
> *Cymbeline, Act V., Sc. V.*

All three of them are desperate: their great guilt,
Like poison given to work a great time after,
Now 'gins to bite the spirits.
> *Tempest, Act III., Sc. III.*

Hubert. The king, I fear, is poison'd by a monk:
I left him almost speechless. *   *   *
Bastard. How did he take it? who did taste to him?
Hubert. A monk, I tell you; a resolved villain,
Whose bowels suddenly burst out: the king
Yet speaks, and, peradventure, may recover.
> *King John, Act V., Sc. VI.*

You shall digest the venom of your spleen,
Though it do split you!
> *Julius Cæsar, Act IV., Sc. III.*

42

If they had swallow'd poison 't would appear
By external swelling : but she looks like sleep.
Antony and Cleopatra, Act V., Sc. II.

K. John. There is so hot a summer in my bosom,
That all my bowels crumble up to dust:
I am a scribbled form, drawn with a pen
Upon a parchment; and against this fire
Do I shrink up.

P. Henry. How fares your majesty ?
K. John. Poison'd,—ill fare; dead, forsook, cast off:
And none of you will bid the winter come,
To thrust his icy fingers in my maw;
Nor let my kingdom's rivers take their course
Through my burn'd bosom; nor entreat the north
To make his bleak winds kiss my parched lips,
And comfort me with cold : I do not ask you much,
I beg cold comfort ; and you are so strait,
And so ingrateful, you deny me that. * * *
Within me is a hell; and there the poison
Is, as a fiend, confin'd to tyrannize
On unreprievable condemned blood.
King John. Act V., Sc. VII.

Within the infant rind of this weak flower
Poison hath residence, and medicine power :
For this, being smelt, with that part cheers each part;
Being tasted, slays all senses with the heart.
Romeo and Juliet, Act II., Sc. III.

Like a poisonous mineral, gnaw my inwards.
Othello, Act II , Sc. I.

I bought an unction of a mountebank,
So mortal, that but dip a knife in it,
Where it draws blood no cataplasm so rare
Collected from all simples that have virtue
Under the moon, can save the thing from death
That is but scratch'd withal.
Hamlet, Act IV., Sc. VII.

A few miscellaneous quotations referring to medical subjects
must here find a place.

The more one sickens the worse at ease he is.
As You Like It, Act III., Sc. II.

43

He fell sick suddenly, and grew so ill
He could not sit his mule.
Henry VIII., Act IV., Sc. II.

——the sun is a most glorious sight,
I've seen him rise full oft, indeed of late
I have set up on purpose all the night,
Which hastens, as physicians say, one's fate ;
And so all ye, who would be in the right
In health and purse, begin your day to date
From day-break, and when coffin'd at fourscore,
Engrave upon the plate you rose at four.
Byron—Don Juan, Canto II., Verse CXL.

So much was our love,
We would not understand what was most fit ;
But, like the owner of a foul disease,
To keep it from divulging, let it feed
Even on the pith of life.
Hamlet, Act IV., Sc. 1.

Diseases desperate grown,
By desperate appliance are reliev'd
Or not at all.
Hamlet, Act IV., Sc. III.

His dissolute disease will scarce obey this medicine.
Merry Wives, Act III., Sc. III.

O vanity of sickness! fierce extremes,
In their continuance, will not feel themselves.
Death, having prey'd upon the outward parts,
Leaves them insensible.
King John, Act V., Sc. VII.

What a catalogue have we here:

Now the rotten diseases of the south, the guts-griping, ruptures, catarrhs,
loads o' gravel i' the back, lethargies, cold palsies, raw eyes, dirt-rotten livers,
wheezing lungs, bladders full of imposthume, sciaticas, lime-kilns i' the palm,
incurable bone-ache, and the rivelled fee-simple of tetter, take and take again
such preposterous discoveries!
Troilus and Cressida, Act V., Sc. 1.

As burning fevers, agues pale and faint,
Life-poisoning pestilence, and frenzies wood,
The marrow-eating sickness, whose attaint
Disorder breeds by heating of the blood :
Surfeits, imposthumes, grief and damn'd despair,
Swear nature's death for framing thee so fair.
Venus and Adonis.

44

How nicely does he describe the decay of man, the second childhood. the wasting away of the organism :

> The sixth age shifts
> Into the lean and slipper'd pantaloon,
> With spectacles on nose and pouch on side ;
> His youthful hose, well sav'd, a world too wide
> For his shrunk shank ; and his big manly voice
> Turning again towards childish treble, pipes
> And whistles in his sound. Last scene of all,
> That ends this strange eventful history,
> Is second childishness, and mere oblivion,
> Sans teeth, sans eyes, sans taste, sans everything.
>
> *As You Like It, Act. II., Sc. VII.*

Again :

Do you set down your name in the scroll of youth, that are written down old with all the characters of age ? Have you not a moist eye? a dry hand ? a yellow cheek ? a white beard ? a decreasing leg? an increasing belly? Is not your voice broken ? your wind short? your chin double? your wit single ? and every part of you blasted with antiquity ; and will you yet call yourself young?

> *Henry IV—2d, Act I., Sc. II.*

The satirical rogue says here, that old men have grey beards; that their faces are wrinkled; their eyes purging thick amber and plum-tree gum ; and that they have a plentiful lack of wit, together with most weak hams.

> *Hamlet, Act II., Sc. II.*

A good leg will fall ; a straight back will stoop; a black beard will turn white ; a curled pate will grow bald ; a fair face will wither; a full eye will wax hollow. * * *

> *Henry V., Act V., Sc. II.*

> Were I hard-favour'd, foul, or wrinkled-old,
> Ill-natur'd, crooked, churlish, harsh in voice,
> O'er worn, despised, rheumatic, and cold,
> Thick-sighted, barren, lean, and lacking juice,
> Then might thou pause. * * *
>
> *Venus and Adonis.*

> Let them die, that age and sullens have ;
> * * * both become the grave.
>
> *Richard II., Act II., Sc. I.*

> Thus, methinks, I hear them speak,
> See, how the Dean begins to break !
> Poor gentleman ! he droops apace !
> You plainly find it in his face.
> That old vertigo in his head

45

Will never leave him, till he's dead.
Besides, his memory decays :
He recollects not what he says :
He can not call his friends to mind ;
Forgets the place where last he dined ;
Plies you with stories o'er and o'er:
He told them fifty times before.
How does he fancy we can sit
To hear his out-of-fashion wit ?
But he takes up with younger folks,
Who for his wine will bear his jokes.
Faith, he must make his stories shorter,
Or change his comrades once a quarter.

Swift—" Death of Dr. Swift."

Thus Swift predicted his own end as early as 1731. History mournfully testifies that his candle burnt out as he anticipated. "Fits of lunacy were succeeded by the *dementia* of old age. For three years he uttered only a few words and broken interjections. He would often attempt to speak, but could not recollect words to express his meaning, upon which he would sigh heavily. Babylon in ruins (to use a *simile* of Addison's), was not a more melancholy spectacle than this wreck of a mighty intellect ! In speechless silence his spirit passed away October 19, 1745." (Chamber's Eng. Lit.)

Manhood declines—age palsies every limb :
He quits the scene—or else the scene quits him ;
Scrapes wealth, o'er each departing penny grieves,
And avarice seizes all ambition leaves ;
Counts cent. per cent., and smiles or vainly frets,
O'er hoards diminish'd by young Hopeful's debts ;
Weighs well and wisely what to sell or buy,
Complete in all life's lessons—but to die ;
Peevish and spiteful, doting, hard to please,
Commending every time, save times like these ;
Crazed, querulous, forsaken, half forgot,
Expires unwept—is buried—let him rot !

Byron—Hints from Horace.

The signs of a probable fatal termination are most beautifully portrayed by Shakespeare. The death of Falstaff can not fail to be regarded by the profession as an excellent description of approaching dissolution.

'A made a finer end, and went away, an it had been any christom child ; 'a parted even just between twelve and one, even at the turning of the tide: for after I saw him fumble with the sheets, and play with flowers, and smile upon his finger's ends, I knew there was but one way; for his nose was as sharp as a pen, and 'a babbled of green fields. * * * 'A bade me lay more clothes on his feet: I put my hand into the bed and felt them, and they were as cold as any stone; then I felt to his knees, and so upwards, and upwards, and all was as cold as any stone.

Henry V., Act II., Sc. III.

Clarence.   Lord! Methought, what pain it was to drown!
What dreadful noise of waters in mine ears!
What ugly sights of death within mine eyes!
   *   *   *   *   *   *   *   *   *

Brakenbury. Had you such leisure in the time of death,
To gaze upon these secrets of the deep?

Clarence.   Methought I had ; for still the envious flood
Kept in my soul and would not let it forth
To seek the empty, vast, and wand'ring air ;
But smother'd it within my panting bulk,
Which almost burst to belch it in the sea.

Richard III., Act I. Sc. IV.

How oft when men are at the point of death,
Have they been merry! which their keepers call
A lightning before death.

Romeo and Juliet, Act V., Sc. III.

Out, alas! she's cold ;
Her blood is settled, and her joints are stiff;
Life and these lips have long been separated :
Death lies on her like an untimely frost
Upon the sweetest flower of all the field.

Romeo and Juliet, Act IV., Sc. V.

Do you notice
How much her grace is alter'd on the sudden?
How long her face is drawn? how pale she looks,
And of an earthy cold! Mark her eyes.
* * * She is going.   *Henry VIII., Act IV., Sc. II.*

Her physician tells me
She hath pursu'd conclusions infinite
Of easy ways to die.   *Antony and Cleopatra, Act V., Sc. II.*

Bid a sick man in sadness make his will:—
A word ill urg'd to one that is so ill.

Romeo and Juliet, Act I., Sc. I.

By his gates of breath
There lies a downy feather, which stirs not:
Did he suspire, that light and weightless down
Perforce must move.   *Henry IV—2d, Act IV., Sc. IV.*

Lend me a looking-glass ;
If that her breath will mist or stain the stone,
Why then she lives.   *King Lear, Act V., Sc. III.*

47

Death, on a solemn night of state,
In all his pomp of terror sate :
The attendants of his gloomy reign,
Diseases dire, a ghastly train !
Crowded the vast court. With hollow tone,
A voice thus thundered from the throne :
" This night our minister we name ;
Let every servant speak his claim ;
Merit shall bear this ebon wand."
All, at the word, stretched forth their hand.
Fever, with burning heat possessed,
Advanced, and for the wand addressed :
" I to the weekly bills appeal ;
Let those express my fervant zeal ;
On every slight occasion near,
With violence I persevere "
Next Gout appears with limping pace,
Pleads how he shifts from place to place :
From head to foot how swift he flies,
And every joint and sinew plies ;
Still working when he seems supprest,
A most tenacious stubborn guest.
A haggard spectre from the crew
Crawls forth, and thus asserts his due :
" 'Tis I who taint the sweetest joy,
And in the shape of love destroy.
My shanks, sunk eyes, and noseless face,
Prove my pretension to the place."
Stone urged his overgrowing force ;
And, next consumption's meagre corse,
With feeble voice that scarce was heard,
Broke with short coughs, his suit preferred :
" Let none object my lingering way :
I gain, like Fabius, by delay ;
Fatigue and weaken every foe
By long attack, secure, though slow."
Plague represents his rapid power,
Who thinned a nation in an hour.
All spoke their claim and hoped the wand.
Now expectation hushed the band,
When thus the monarch from the throne :
" Merit was ever modest known.
What ! no physician speak his right?
None here ! but fees their toil requite.
Let, then, Intemperance take the wand,
Who fills with gold their zealous hand.
You, Fever, Gout, and all the rest—
Whom wary men as foes detest—
Forego your claim. No more pretend
Intemperance is esteemed a friend ;
He shares their mirth, their social joys,
And as a courted guest destroys.
The charge on him must justly fall,
Who finds employment for you all "   *Gay—" Court of Death."*

48

PART III.

SURGERY.

Shakespeare paid much more attention to the practice of medicine and obstetrics than to surgery. Perhaps the cause of this was that at that time surgery had not reached its present perfection. A more probable reason is that his son-in-law, Dr. John Hall, may not have been a surgeon.

> *Iago.* What, are you hurt, lieutenant?
> *Cas.* Ay, past all surgery.
>> *Othello, Act II., Sc. III.*

Can honour set a leg? No. Or an arm? No. Or take away the grief of a wound? No. Honour hath no skill in surgery then? No.
>> *Henry IV., Act V., Sc. I.*

With the help of a surgeon he might yet recover.
>> *Midsummer Night's Dream, Act V., Sc. I.*

> Let me have surgeons;
> I am cut to the brains.
>> *King Lear, Act IV., Sc. VI.*

The king himself hath a heavy reckoning to make when all those legs, and arms, and heads, chopped off in a battle, shall join together at the latter day, and cry all——We died at such a place; some swearing, some crying for a surgeon, some, upon their wives left poor behind them.
>> *Henry V., Act IV., Sc. I.*

> *Patr.* Who keeps the tent now?
> *Ther.* The surgeon's box, or the patient's wound.
>> *Troilus and Cressida, Act V., Sc. I.*

Give physic to the sick, ease to the pain'd:
The poor, lame, blind, halt, creep, cry out for thee.
>> *Lucrece.*

What opposite discoveries we have seen!
(Signs of true genius, and of empty pockets;)
One makes new noses, one a guillotine,
One breaks your bones, one sets them in their sockets.
>> *Byron—Don Juan, Canto I., Verse CXXIX.*

49

The lawyer's brief is like the surgeon's knife
Dissecting the whole inside of a question,
And with it all the process of digestion.

> *Byron—Don Juan, Canto X., Verse XIV.*

All feel the ill, yet shun the cure.
Can sense this paradox endure?

> *Swift.*

Syphilis is frequently referred to, and he represents several of his characters as having it; among them Falstaff and Dame Quickly.

Lysimachus to keeper of a bawdy house:

Have you that a man may deal withal and defy the surgeon?

> *Pericles, Act IV., Sc. VI.*

You help to make the diseases, Doll:
We catch of you, Doll, we catch of you.

> *Henry IV—2d, Act II., Sc. IV.*

Boult. Do you know the French knight that cowers i' the hams? * * *
Bawd. As for him he brought his disease hither.

> *Pericles, Act IV., Sc. II.*

Doth fortune play the huswife with me now?
News have I, that my Nell is dead i' the spital
Of malady of France.

> *Henry V., Act V., Sc. I.*

In this sty, where, since I came,
Diseases have been sold dearer than physic.

> *Pericles, Act IV., Sc. VI.*

With tomboys, * * * with diseas'd ventures,
That play with all infirmities for gold,
Which rottenness can lend nature!
Such boil'd stuff
As well might poison poison!

> *Cymbeline, Act I., Sc. VI.*

I have purchased as many diseases under her roof as come to * * * * three thousand dollars a year.

> *Measure for Measure, Act I, Sc. II.*

Nor did not with unbashful forehead woo
The means of weakness and debility.

> *As You Like It, Act II., Sc. III.*

If we two be one, and thou play false,
I do digest the poison of thy flesh.

> *Comedy of Errors, Act II., Sc. II.*

50

> Consumptions sow
> In *hollow bones of* men; strike their *sharp shins,*
> And mar men's spurring. *Crack the* lawyer's *voice,*
> That he may never more false title plead,
> Nor *sound* his quillets *shrilly:* hoar the flamen,
> That scolds against the quality of flesh,
> And not believes himself: *down with the nose,*
> *Down with it flat; take the bridge quite away,*
> Of him that, his particular to foresee,
> *Smells from the general weal: make curl'd pate* ruffians *bald;*
> And let the unscarr'd braggarts of the war
> *Derive* some *pain* from you.
> *Timon of Athens, Act IV., Sc. III.*

The symptoms of secondary and tertiary syphilis are accurately expressed in this curse of Timon's. Leprosy is referred to in the sentence " hoar the flamen," or in other words, make white the priest. Shakespeare here shows a very fine point by using these most dreaded of all diseases : leprosy, syphilis, and consumption—maladies that are hereditary, incurable, and contagious. They are certainly lasting,as he wishes the curse to be.

> A pox on 't!

A common expression scattered through many of his plays.

> A man can no more separate age and covetousness than he can part young limbs and lechery; but the gout galls the one, and the pox pinches the other.
> *Henry IV—2d, Act I., Sc. II.*

> I'faith, if he be not rotten before he die (as we have many pocky corses now-a-days, that will scarce hold the laying in), he will last you some eight year or nine year.
> *Hamlet, Act V., Sc. I.*

> She hath eaten up all her beef, and is herself in the tub.
> *Measure for Measure, Act III., Sc. II.*

> To the spital go,
> And from the powdering-tub of infamy
> Fetch forth the lazar-kite of Cressid's kind,
> Doll Tearsheet she by name. *Henry V., Act II., Sc. I.*

> Be a whore still : * * * *
> Give them diseases. * * *
> * * * * Season the slaves
> For tubs and baths; bring down rose-cheeked youth
> To the tub-fast, and the diet.
> *Timon of Athens, Act IV., Sc. III.*

51

Dr. Macdonnell, of Canada, has thrown much light on these quotations in his works on Syphilis. He says: "It appears to have been the custom to prescribe for syphilitic patients, in addition to inunction, a prolonged diaphoresis and a very low diet. On the continent the patient was placed in a cave, oven, or dungeon, and Wiseman says it was the custom in England to use a tub for this purpose."

In the foot-note to the passage in Johnson & Steven's edition of Shakespeare's works the following quotations from old plays are given :

> "——you had better match a ruin'd bawd,
> One ten times cur'd by sweating and the tub."
>
> *Jaspar Maines*, 1639.

Again, in the *Family of Love*, (1608), a doctor says :

" O for one of the hoops of my Cornelius' tub, I shall burst myself with laughing else."

In *Monsieur d' Olive*, (1606) :

"Our embassage is into France, there may be employment for thee: Hast thou a tub?"

> She, whom the spital-house, and ulcerous sores
> Would cast the gorge at, this embalms and spices
> To the April day again.
>
> *Timon of Athens, Act IV., Sc. III.*

> 'Tis I who taint the sweetest joy,
> And in the shape of love destroy.
> My shanks, sunk eyes, and noseless face,
> Prove my pretension to the place.
>
> *Gay.*

> Pox take him and his wit.
>
> *Swift.*

> Constant to nought—save hazard and a whore,
> Yet cursing both - for both have made him sore ;
> Unread—unless, since books beguile disease,
> The pox becomes his passage to degrees.
>
> *Byron—Hints from Horace.*

> I said small-pox had gone out of late ;
> Perhaps it will be followed by the great.
> 'Tis said the great came from America ;
> Perhaps it may set out on its return,—
> The population there so spreads, they say,
> 'Tis grown high time to thin it in its turn,
> With war, or plague, or famine, any way,
> So that civilization they may learn ;
> And which in ravage the more loathsome evil is—
> Their real lues, or our pseudo-syphilis?
>
> *Byron—Don Juan, Canto I., Verse CXXX.*

> He'll feel the weight of it many a day.
>
> *Cowley.*

A little attention is paid to diseases of the eye, thus in Winter's Tale :

> Wishing all eyes
> Blind with the pin and web, but theirs, theirs only,
> That would unseen be wicked.

Act I., Sc. II.

Commentators have the thought that Shakespeare wished to express the idea of cataract by the term pin and web—this is, without doubt, a mistake ; he did not intend to make lovers so cruel that they should desire to deprive every one else of sight. Pin and web (being a varicose excrescence of the conjunctiva, sometimes to such an extent as to totally prevent vision), was meant to express a veil, or in other words, the eyelid.

> Must you with hot irons burn out both mine eyes ?
> * * * * * * * * *
> O heaven ! that there were but a mote in yours,
> A grain, a dust, a gnat, a wandering hair,
> Any 'annoyance in that precious sense !
> Then, feeling what small things are boist'rous there,
> Your vile intent must needs seem horrible.

King John, Act IV., Sc. I.

The term "sand-blind" was meant to express a dimness of sight, as if sand had been thrown in the eyes.

Launcelot. O heavens, this is my true-begotten father! who, being more than sand-blind, high-gravel blind, knows me not.
* * * * * * * * *

Gobbo. Alack, sir, I am sand-blind, I know you not.

Merchant of Venice, Act II., Sc. II.

> I remember thine eyes well enough
> Dost thou squiny at me ?

King Lear, Act. IV., Sc. II.

> He gives the web and the pin, squints the eye, and makes the hare-lip.

King Lear, Act III., Sc. IV.

> Thou green sarcenet flap for a sore eye.

Troilus and Cressida, Act V., Sc. I.

> A merry, cock-eyed, curious looking sprite.
>
> *Byron—Vision of Judgment.*

> To no one muse does she her glance confine,
> But has an eye, at once, to all the nine.
>
> *Tom Moore.*

The subject of wounds has received frequent mention.

A scratch, a scratch; marry, 'tis enough; * * * go, villain, fetch a surgeon. * * * 'Tis not deep as a well, nor as wide as a church door; but
'tis enough, * * * ask for me to-morrow, and you shall find me a grave
man.

Romeo and Juliet, Act III., Sc. I.

Have by some surgeon * * *
To stop his wounds lest he do bleed to death.

Merchant of Venice, Act IV., Sc. I.

For the love of God, a surgeon! send one presently to Sir Toby. * * *
H'as broke my head across, and has given Sir Toby a bloody coxcomb too:
for the love of God your help!

Twelfth Night, Act V., Sc. I.

Romeo. Your plantain leaf is excellent for that.
Benvolio. For what, I pray thee?
Romeo. For thy broken shin.

Romeo and Juliet, Act I., Sc. II.

Moth. A wonder, master; here's a Costard broken in a shin.
Armado. Some enigma, some riddle: come,—thy *l'envoy;* begin.
Costard. No egma, no riddle, no *l'envoy;* no salve ih the male, sir; O
      sir, plantain, a plain plantain; * * * no salve, sir, but a
      plantain!   *Love's Labour's Lost, Act III., Sc. I.*

The sovereign'st thing on earth
Was parmaceti, for an inward bruise.

Henry IV., Act I., Sc. III.

I do beseech your majesty, may salve
The long-grown wounds of my intemperance.

Henry IV., Act III., Sc. II.

Let us hence, my sovereign, to provide
A salve for any sore that may betide.

Henry VI—3d, Act. IV., Sc. VI.

Here is a letter, lady;
The paper as the body of my friend,
And every word in it a gaping wound,
Issuing life-blood.   *Merchant of Venice, Act III., Sc. II.*

He jests at scars, that never felt a wound.

Romeo and Juliet, Act II., Sc. II.

Dercetas.   This is his sword;
        I robb'd his wound of it. * * *
Cæsar.   * * * We do lance
        Diseases in our bodies.

Antony and Cleopatra, Act V., Sc. I.

Men. Where is he wounded?

Vol. I' the shoulder and i' the left arm :
There will be large cicatrices to show the people.

> *Coriolanus, Act II., Sc. I.*

What wound did ever heal but by degrees?

> *Othello, Act II., Sc. III.*

To see the salve doth make the wound ache more.

> *Lucrece.*

Scratch thee but with a pin, and there remains
Some scar of it.

> *As You Like It, Act III., Sc. V.*

The new-heal'd wound * * * should break out,
Which would be so much the more dangerous.

> *Richard III., Act II.. Sc II.*

I shall desire you of more acquaintance, good master cobweb. If I cut my finger, I shall make bold with you.

> *Midsummer Night's Dream, Act III., Sc. I.*

I'll fetch some flax, and whites of eggs
To apply to 's bleeding face.

> *King Lear, Act III., Sc. VII.*

Go, get a white of an egg and a little flax, and close the breach of the head; it is the most conducible thing that can be.

> *Ben Jonson — " The Case is Altered." Act II., Sc. IV,*

One's hip he slash'd, and split the other's shoulder,
And drove them with their brutal yells to seek
If there might be chirurgeons who could solder
The wounds they richly merited.

> *Byron—Don Juan, Canto VIII., Verse XCIV.*

Many surgical subjects receive but little attention from him.

Ber. What is it, my good lord, the king languishes of?

Laf. A fistula, my lord.

> *All's Well, Act I., Sc. I.*

Fal. Why, sirs, I am almost out at heels.

Pist. Why, then, let kibes ensue.

> *Merry Wives, Act I., Sc. III.*

The age is grown so picked, that the toe of the peasant comes so near the heel of the courtier, he galls his kibe.

> *Hamlet, Act V., Sc. I.*

If it were a kibe
'Twould put me to my slipper.

> *Tempest, Act II., Sc. I.*

55

If a man's brains were in 's heels, were 't not in danger of kibes?
King Lear, Act I., Sc. V.

Does your worship mean to geld and splay all the youth of the city?
Measure for Measure, Act II., Sc. I.

Thou hast drawn my shoulder out of joint.
Henry IV—2d, Act V., Sc. IV.

Were 't my fitness
To let these hands obey my blood,
They are apt enough to dislocate and tear
Thy flesh and bones :—howe'er thou art a fiend,
A woman's shape doth shield thee.
King Lear, Act IV., Sc. II.

Charles in a moment threw him, and broke three of his ribs, * * * *
there is little hope of life in him.
As You Like It, Act I., Sc. II.

It is the first time that ever I heard breaking of ribs was sport for ladies.
As You Like It, Act I., Sc. II.

On her left breast
A mole cinque-spotted, like the crimson drops
I' the bottom of a cowslip.
Cymbeline, Act II., Sc. II.

Under her breast
(Worthy the pressing) lies a mole, right proud
Of that most delicate lodging.
Cymbeline, Act II., Sc. IV.

If thou wert * * * *
Lame, foolish, crooked, swart, prodigious,
Patch'd with foul moles and eye offending marks,
I would not care. * * *
King John, Act III., Sc. I.

In case of a recent burn it was the custom to place the part near the fire, thus upholding the old homœopathic doctrine that what hurts will cure.

And falsehood falsehood cures; as fire cools fire
Within the scorched veins of one new burn'd.
King John, Act III., Sc. I.

One fire drives out one fire; one nail, one nail;
Rights by rights founder, strength by strengths do fail.
Coriolanus, Act IV., Sc. VII.

One fire burns out another's burning,
One pain is lessen'd by another's anguish.
Romeo and Juliet, Act I., Sc. II.

Even as one heat another heat expels,
Or as one nail by strength drives out another,
So the remembrance of my former love
Is by a newer object quite forgotten.
Two Gentlemen of Verona, Act II., Sc IV.

I must not break my back to heal his finger.
Timon of Athens, Act II., Sc. I.

That bottled spider, that foul, bunch-back'd toad.
Richard III., Act IV., Sc. IV.

Where's that valiant crook-back prodigy?
Henry VI—3d, Act I., Sc. IV.

Ladies, that have their toes
Unplagu'd with corns, will have a bout with you. * *
* * * Which of you all
Will now deny to dance? she that makes dainty,
She, I'll swear, hath corns.
Romeo and Juliet, Act I., Sc. V.

Strangely-visited people,
All swoln and ulcerous, pitiful to the eye,
The mere despair of surgery.
Macbeth, Act IV., Sc. III.

Fell sorrow's tooth doth never rankle more
Than when it bites but lanceth not the sore.
Richard II., Act I., Sc. III.

You rub the sore,
When you should bring the plaster.
Tempest, Act II., Sc. I.

It will but skin and film the ulcerous place.
Hamlet, Act III., Sc. IV.

Men.   The service of the foot
Being once gangren'd is not then respected
For what before it was.
Bru. Pursue him to his house, and pluck him thence,
Lest his infection, being of catching nature,
Spread further.
Coriolanus, Act III., Sc. I.

Sic. He's a disease that must be cut away.
Men. O he's a limb that has but a disease;
Moral, to cut it off; to cure it easy.

Coriolanus, Act III., Sc. I.

Falstaff. Boy, tell him I am deaf.
Page. You must speak louder, my master is deaf.
* * * * * * * *
Falstaff. * * * it is a kind of deafness.
Ch. Just. I think you are fallen into the disease; for you hear not what I
say to you, * * * and I care not if I do become your physician.
Falstaff. * * * I should be your patient to follow your prescriptions,
the wise may make some dram of a scruple, or, indeed, a scruple
itself.

Henry IV—2d, Act I., Sc. II.

The surgery described in Titus Andronicus is, of course, impossible.

With gaping mouth.

Spenser.

Madame scolded one day so long,
She sudden lost all use of tongue.
The doctor came—with hem and haw,
Pronounced the affection a lock'd jaw.

———

Let firm, well-hammered soles protect thy feet
Through freezing snows, and rains, and soaking sleet.
Should the big last extend the shoe too wide,
Each stone will wrench the unwary step aside ;
The sudden turn may stretch the swelling vein,
The cracking joint unhinge, or ankle sprain ;
And when too short the modish shoes are worn,
You'll judge the seasons by your shooting corn

Gay.

Leeches stick, nor quit the bleeding wound,
Till off they drop with skinfuls to the ground.

Swift.

Think of the thunderer's falling down below
Carotid-artery-cutting Castlereagh !
Alas ! that glory should be chill'd by snow !

Byron—Don Juan, Canto X, Verse LIX.

.The surgeon had his instruments and bled
Pedrillo, and so gently ebb'd his breath,
You hardly could perceive when he was dead.
* * * * * * * *
And first a little crucifix he kissed,
And then held out his jugular and wrist.

Byron—Don Juan, Canto II., Verse LXXVI.

PART IV.

OBSTETRICS.

Obstetrics was Shakespeare's favorite branch of the profession, and he has not been at all sparing in reference to it. Under this head will be included many topics which could more properly be placed in the chapter on physiology, but it is thought better to have such intimate subjects classed together. They have been arranged in the order of their natural occurrence.

> *Capulet.* My child is yet a stranger in the world,
> She hath not seen the change of fourteen years;
> Let two more summers wither in their pride,
> Ere we may think her ripe to be a bride.
> *Paris.* Younger than she are happy mothers made.
> *Capulet* And too soon marr'd are those so early made.
> *Romeo and Juliet, Act I., Sc. II.*

> Well, think of marriage now; younger than you,
> Here in Verona, ladies of esteem,
> Are made already mothers: by my count,
> I was your mother much upon these years
> That you are now a maid.
> *Romeo and Juliet, Act I., Sc. III.*

In the old poem Juliet's age is put down as sixteen; in Paynter's novel she is said to be eighteen. Shakespeare, however, makes her fourteen, but who ever imagines her of these tender years while enjoying the play? It seems absurd to think of her as being less than twenty or twenty-two until we recollect that she grew and developed into early womanhood under the sun of an Italian clime. The wonderful development of the girls of Italy can easily be seen in the Eternal city. Taking a stroll down to the Spanish staircase which is daily filled with Roman models lazily awaiting the engagements of the artists, or a walk on the Corso, or around the Theatre of Marcellus, convinces one at once that Shakespeare's Juliet, young as she is, is not over-

drawn, and that the Italian girl of fourteen is indeed fully
"ripe to be a bride."

'Tis a sad thing, I can not choose but say,
And all the fault of that indecent sun
Who can not leave alone our helpless clay,
But will keep baking, broiling, burning on,
That, howsoever people fast and pray,
The flesh is frail and so the soul's undone :
What men call gallantry, and gods adultery,
Is much more common where the climate's sultry.
Byron—Don Juan, Canto I., Verse LXIII.

Shakespeare has hinted several times that it was a common
occurrence for girls of this "sun-burnt nation" to be mothers
at the age of fourteen. Paris assures Juliet's father that "younger
than she are happy mothers made," and Lady Capulet, in her
conversation with her daughter, alludes to the fact that she was
her mother when she was but thirteen. She also echoes Paris
in saying :

Younger than you
Here in Verona, ladies of esteem,
Are made already mothers.

Another reference is found in Winter's Tale :

If this prove true, they'll pay for it: by mine honour,
I'll geld 'em all; fourteen they shall not see,
To bring false generations. *Act II., Sc. I.*

Perhaps Byron had a better idea of this climatic effect than
any other poet. He has frequently written of it; indeed, it
forms the foundation of some of his poems.

Wedded she was some years, and to a man
Of fifty and such husbands are in-plenty ;
And yet, I think, instead of such a *one*,
'Twere better to have two of five and twenty,
Especially in countries near the sun.
Byron—Don Juan, Canto I., Verse LXII.

It was upon a day, a summer's day ;
Summer 's indeed a very dangerous season,
And so is spring about the end of May ;
The sun, no doubt, is the prevailing reason.
Byron—Don Juan, Canto I., Verse CII.

Haidee was nature's bride, and knew not this ;
Haidee was passion's child, born where the sun
Showers triple light. and scorches even the kiss
Of his gazelle-eyed daughters.
Byron—Don Juan, Canto II., Verse CCII.

The Turks do well to shut—at least sometimes—
The women up—because, in sad reality,
Their chastity in these unhappy climes
Is not a thing of that astringent quality,
Which in the north prevents precocious crimes.
Byron—Don Juan, Canto V., Verse CLVII.

Few short years make wondrous alterations,
Particularly among sun-burnt nations.
Byron—Don Juan, Canto I., Verse LXIX.

Our English maids are long to woo,
And frigid even in possession;
And if their charms be fair to view,
Their lips are slow at love's confession:
But born beneath a brighter sun,
For love ordain'd the Spanish maid is
And who when fondly, fairly won,—
Enchants you like the girl of Cadiz?
*   *   *   *   *   *   *
In each her charms the heart must move
Of all who venture to behold her;
Then let not maids less fair reprove
Because her bosom is not colder:
Through many a clime 'tis mine to roam
Where many a soft and melting maid is,
But none abroad and few at home
May match the dark-eyed girl of Cadiz.
Byron—Poems.

What a beautiful comparison Shakespeare has made between
the virgin and flowers.

I would I had some flowers o' the spring, that might
Become your time of day; and yours, and yours,
That wear upon your virgin branches yet
Your maidenheads growing * * *
* * * * pale primroses,
That die unmarried, ere they can behold
Bright Phœbus in his strength,—a malady
Most incident to maids.
Winter's Tale, Act IV., Sc. III.

Fair Hermia, question your desires,
Know of your youth, examine well your blood,
Whether, if you yield not to your father's choice,
You can endure the livery of a nun;
For aye to be in shady cloister mew'd
To live a barren sister all your life,
Chanting faint hymns to the cold fruitless moon.
Thrice blessed they that master so their blood,
To undergo such maiden pilgrimage;

But earthly happier is the rose distill'd,
Than that, which, withering on the virgin thorn,
Grows, lives, and dies in single blessedness.
Midsummer Night's Dream, Act I., Sc. I.

Fecundation is not overlooked, and Shakespeare shows his knowledge of the fact that the penis is merely the spout or funnel by which the semen is conveyed to the uterus, and aptly compares the womb to a bottle, which in his time gradually tapered toward the neck. The word tundish is an old Warwickshire name for a funnel.

Duke. Why should he die, sir?
Lucio. Why? For filling a bottle with a tun-dish.
Measure for Measure, Act III., Sc. II.

Thou shalt not die: die for adultery! No:
The wren goes to 't, and the small gilded fly
Does lecher in my sight.
Let copulation thrive for Gloster's bastard son
Was kinder to his father than my daughters
Got 'tween lawful sheets.
King Lear, Act IV., Sc. VI.

Hymen hath brought the bride to bed,
Where, by the loss of maidenhead,
A babe is moulded.
Pericles, Gow to Act III.

Crack nature's moulds, all germens spill at once,
That make ungrateful man.
King Lear, Act III., Sc. II.

Q. Eliz. But thou didst kill my children.
K. Rich. But in your daughter's womb I'll bury them;
Where, in that nest of spicery, they shall breed
Selves of themselves, to your recomforture.
Richard III., Act IV., Sc. IV.

Your brother and his lover have embrac'd:
As those that feed grow full; as blossoming time,
That from the seedness the bare fallow brings
To teeming foison, even so her plenteous womb
Expresseth his full tilth and husbandry.
Measure for Measure, Act I., Sc. IV.

Hear, nature, hear; dear goddess hear!
Suspend thy purpose, if thou didst intend

To make this creature fruitful!
Into her womb convey sterility!
Dry up in her the organs of increase; ·
And from her derogate body never spring
A babe to honour her! If she must teem,
Create her child of spleen; that it may live,
And be a thwart disnatur'd torment to her!
Let it stamp wrinkles in her brow of youth;
With cadent tears fret channels in her cheeks;
Turn all her mother's pains and benefits
To laughter and contempt: that she may feel
How sharper than a serpent's tooth it is
To have a thankless child!
King Lear, Act I., Sc. IV.

The production of either sex at will agitated the minds of physiologists to a considerable extent during Shakespeare's time. Indeed he seems to have held an ancient theory that the more vigorous of the parents produced the opposite sex. Dr. Robert, of Paris, in his paper entitled *Megalanthropogenesis*, somewhat followed up this theory and maintained that "the race of men of genius might be perpetuated by uniting them to better physically developed women having clever minds," which, according to his theory, would, of course, result in nothing but male children.

Bring forth men-children only!
For thy undaunted mettle should compose
Nothing but males.
Macbeth, Act I., Sc. VII.

For men's sake, the authors of these women;
Or women's sake, by whom we men are men.
Love's Labour's Lost, Act IV., Sc. III.

Be advis'd, fair maid:.
To you your father should be as a god;
One that compos'd your beauties; yea, and one
To whom you are but as a form in wax,
By him imprinted, and within his power
To leave the figure, or disfigure it.
Midsummer Night's Dream, Act I, Sc. I.

The child would therefore resemble the parent of opposite sex.

Nurse to Henry VIII:

> 'Tis a girl * * * as like you
> As cherry is to cherry.
>
> *Act V., Sc. I.*

Paulina pleading to Leontes on the birth of a daughter to his wife Hermione:

> Behold, my lords,
> Although the print be little, the whole matter .
> And copy of the father,—eye, nose, lip;
> The trick of 's frown ; his forehead ; nay, the valley,
> The pretty dimples of his chin and cheek ; his smiles;
> The very mould and frame of hand, nail, finger.
>
> *Winter's Tale, Act II., Sc. III.*

It is a very old opinion that the mental state of parents during coition influenced to a certain extent the mental activity of
the offspring. Bastards were supposed to excel in this respect
on account of the mental excitement during the intercourse
from which they took their origin. Burton held this view in his
"Anatomy of Melancholy," and, after reading King Lear, we
know that Shakespeare also held it.

> *Edmund.*   Why brand they us
> With base? with baseness? bastardy? base? base?
> Who in the lusty stealth of nature take
> More composition and fierce quality
> Than doth, within a dull, stale, tired bed
> Go to the creating a whole tribe of fobs,
> Got 'tween sleep and wake.
>
> *Act. I., Sc. II.*

His allusions to pregnancy are many.

> He knows himself my bed he hath defil'd ;
> And at that time he got his wife with child :
> Dead though she be, she feels her young one kick ;
> So there's my riddle, One that's dead is *quick.*
>
> *All's Well, Act V., Sc. III.*

> She is gone ; she is two month on her way. * *
> She 's quick ; the child brags in her belly already.
>
> *Love's Labour's Lost, Act V., Sc. II.*

A mistake of ten weeks is truly a bad one; quickening generally being experienced *four and a half months* after impregna-
tion.

I am with child, * * * *
Murder not, then, the fruit within my womb.
Henry VI., Act V., Sc. IV.

She died, but not alone; she held within
A second principle of life, which might
Have dawn'd a fair and sinless child of sin :
But closed its little being without light,
And went down to the grave unborn, wherein
Blossom and bough lie wither'd with one blight.
Byron—Don Juan, Canto IV., Verse LXX.

This blue ey'd hag was hither brought with child.
Tempest, Act I., Sc. II.

If myself might be his judge,
He should receive his punishment in thanks:
He hath got his friend with child.
Measure for Measure, Act I, Sc. IV.

I shall answer that * * * better than you can the getting up of the negro's belly; the moor is with child.
Merchant of Venice, Act III., Sc. V.

I would there were no age between ten, and three and twenty, or that youth would sleep out the rest; for there is nothing in the between but getting wenches with child, wronging the ancientry, stealing, fighting. * * *
Winter's Tale, Act III., Sc III.

He was whipped for getting the shrieve's fool with child; a dumb innocent that could not say him nay.
All's Well, Act IV., Sc. III.

Let wives with child
Pray that their burthens may not fall this day.
King John, Act III., Sc. I.

Shakespeare knew of the importance of pregnant women, being particularly careful that nothing should excite them.

I the rather wean me from despair,
For love of Edward's offspring in my womb:
This is it that makes me bridle passion,
And bear with mildness my misfortune's cross;
Ay, ay, for this I draw in many a tear,
And stop the rising of blood-sucking sighs,
Lest with my sighs or tears I blast or drown
King Edward's fruit, true heir to the English crown.
Henry VI—3d, Act IV., Sc. IV.

The longings or desires of pregnant women are very nicely shown in Measure for Measure :

> She came in great with child, and longing for stewed prunes.
> > *Act II., Sc. I.*

This mistress Elbow, being as I say, with child, and being great bellied, and longing, as I said, for prunes. * * *
> *Measure for Measure, Act II., Sc. I.*

> From whom my absence was not six months old,
> Before herself (almost at fainting under
> The pleasing punishment that women bear)
> Had made provision for her following me.
> > *Comedy of Errors, Act I., Sc. I.*

> The queen rounds apace. * * *
> * * * She is spread of late
> Into a goodly bulk.
> > *Winter's Tale, Act II., Sc. I.*

> The queen, your mother, rounds apace: we shall
> Present our services to a fine new prince
> One of these days.
> > *Winter's Tale, Act II., Sc. I.*

She grew round-wombed, and had a son for her cradle ere she had a husband for her bed.
> *King Lear, Act I., Sc. I.*

> Great-bellied women,
> That had not half a week to go, like rams
> In the old time of war, would shake the press
> And make 'em reel before 'em.
> > *Henry VIII., Act IV., Sc. I.*

Parturition is referred to in many instances.

> Lucina, O
> Divinest patroness, and midwife gentle
> To those that cry by night, convey thy deity
> Aboard our dancing boat; make swift the pangs
> Of my queen's travails!
> > *Pericles, Act III., Sc. I.*

> What shall be done with groaning Juliet?
> She's very near her hour.
> > *Measure for Measure, Act II., Sc. II.*

Come, let us go, and pray to all the gods
For our beloved mother in her pains. .

Titus Andronicus, Act IV., Sc. II.

The lady shrieks, and well-a-near
Doth fall in travail with her fear.

Pericles, Gow to Act III.

She is deliver'd, lords,—she is deliver'd.
I mean, she is brought a-bed.

Titus Andronicus, Act IV., Sc. II.

The queen's in labour,
They say, in great extremity ; and fear'd
She'll with the labour end.

Henry VIII., Act V., Sc. I.

The queen's in labour. * * * Her sufferance made
Almost each pang a death.

Henry VIII , Act V., Sc. I.

Finger of birth-strangled babe
Ditch-deliver'd by a drab. * * *

Macbeth, Act IV., Sc. I.

You ne'er oppressed me with a mother's groan,
Yet I express to you a mother's care.

All's Well, Act I., Sc. I.

History records the fact that the Duke of Gloucester, afterwards Richard III., was born with teeth, uneven shoulders, one leg shorter than the other, deformed back, with a clump of hair on it. These facts Shakespeare never forgot, and continually harps on them.

Thy mother felt more than a mother's pain,
And yet brought forth less than a mother's hope;
To wit, an indigest deformed lump,
Not like the fruit of such a goodly tree.
Teeth hadst thou in thy head when thou wast born,
To signify, thou cam'st to bite the world.

Henry VI—3d., Act V., Sc. VI.

I have often heard my mother say
I came into the world with my legs forward :
Had I not reason, think ye, to make haste,
And seek their ruin that usurp'd our right?
The midwife wonder'd and the women cried,
O, Jesus bless us, he is born with teeth !

And so I was, which plainly signified
That I should snarl, and bite, and play the dog.
Henry VI—3d., Act V., Sc. VI.

Love forswore me in my mother's womb:
And, for I should not deal in her soft laws,
She did corrupt frail nature with some bribe
To shrink mine arm up like a wither'd shrub;
To make an envious monntain on my back,
Where sits deformity to mock my body;
To shape my legs of an unequal size;
To disproportion me in every part,
Like to a chaos, or an unlick'd bear-whelp
That carries no impression like the dam.
Henry VI—3d, Act III., Sc. II.

The term "unlick'd bear-whelp," in the last quotation, refers
to an old notion existing before Shakespeare's time: that the
bear brings forth masses of animated flesh, having no resem-
blance whatever to her, and that she then licks this shapeless
lump into a cub. There is a thread of truth running through
this idea, as will be seen by the following extract taken by Dyer
from "Arcana Microcosmi," by Alexander Ross: "Bears bring
forth their young deformed and misshapen, by reason of the
thick membrane in which they are wrapped, that is covered over
with a mucous matter. This, he says, the dam contracts in the
winter-time, by lying in hollow caves without motion, so that to
the eye the cub appears like an unformed lump. The above
mucilage is afterwards licked away by the dam, and the membrane broken, whereby that which before seemed to be unformed
appears now in its right shape." Ross holds that this was well
known by the ancients and that they entertained no other idea
in regard to it.

Hence, heap of wrath, foul indigested lump,
As crooked in thy manners as thy shape!
Henry VI—2d, Act V., Sc. I.

I, that am curtail'd of this fair proportion,
Cheated of feature by dissembling nature,
Deform'd, unfinish'd, sent before my time
Into this breathing world, scarce half made up,
And that so lamely and unfashionable,

That dogs bark at me as I halt by them ;
Why I. * * * since I cannot prove a lover,
I am determined to prove a villain.
Richard III., Act I., Sc. I.

Marry, they say my uncle grew so fast
That he could gnaw a crust at two hours old ;
'Twas full two years ere I could get a tooth.
Richard III., Act II., Sc. IV.

Thou elvish-mark'd, abortive, rooting hog!
Thou that wast seal'd in thy nativity
The slave of nature and the son of hell!
Thou slander of thy mother's heavy womb!
Thou loathed issue of thy father's loins!
Richard III., Act I., Sc. III.

Art thou so hasty ? I have stay'd for thee,
God knows, in anguish, pain and agony.
* * * A grievous burden was thy birth to me.
Richard III., Act IV., Sc. IV.

From forth the kennel of thy womb hath crept
A hell-hound that doth hunt us all to death :
That dog, that had his teeth before his eyes.
Richard III., Act IV., Sc. IV.

A few other quotations referring to labor are here found.

By her he had two children at one birth.
Henry VI—2d, Act IV., Sc. II.

A terrible child-bed hast thou had, my dear;
No light, no fire.
Pericles, Act III., Sc. I.

At sea, in child-bed died she, but brought forth
A maid-child called Marina.
Pericles, Act V., Sc. III.

The child-bed privilege denied, which 'longs
To women of all fashion ;—lastly, hurried
Here to this place, i' the open air, before
I have got strength of limit.
Winter's Tale, Act III., Sc. II.

Alas ! worlds fall—and woman since she fell'd
The world (as, since that history, less polite
Than true, hath been a creed so strictly held)
Has not yet given up the practice quite.
Poor thing of usages ! coerced, compell'd,

Victim when wrong, and martyr oft when right,
Condemn'd to child-bed, as men for their sins,
Have shaving too entail'd upon their chins,—
A daily plague, which, in the aggregate,
May average on the whole with parturition.
But as to women who can penetrate
The real sufferings of their she condition?
Man's very sympathy with their estate
Has much of selfishness and more suspicion.
Their love, their virtue, beauty, education,
But form good housekeepers to breed a nation.

Byron—Don Juan, Canto XIV., Verse XXIII.

They are as children but one step below,
Even of your mettle, of your very blood;
Of all one pain, save for a night of groans
Endur'd of her, for whom you bid like sorrow.

Richard III., Act IV., Sc. IV.

Would I had died a maid,
And never seen thee, never borne thee son,
Seeing thou hast prov'd so unnatural a father!
Hath he deserv'd to lose his birthright thus?
Hadst thou but lov'd him half so well as I,
Or felt that pain which I did for him once,
Or nourish'd him, as I did with my blood.
* * * * * * * *

Henry VI—3d, Act I., Sc. I.

He is your brother, lords; sensibly fed
Of that self-blood that first gave life to you;
And from that womb where you imprison'd were,
He is enfranchised and come to light.

Titus Andronicus, Act IV., Sc. II.

The child was prisoner to the womb, and is
By law and process of great Nature, thence
Freed and enfranchis'd.

Winter's Tale, Act II., Sc. II.

She said, no shepherd sought her side,
No hunter's hand her snood untied,
Yet ne'er again to braid her hair
The virgin snood did Alice wear;
Gone was her maiden glee and sport,
Her maiden girdle all too short,
Nor sought she, from that fatal night,
Or holy church or blessed rite,
But lock'd her secret in her breast,
And died in travail unconfess'd.

Scott—Lady of the Lake, Canto III., Verse V.

My princely father then had wars in France;
And by true computation of the time,
Found that the issue was not his begot.
Richard III., Act III., Sc. V.

Worse than a slavish wipe, or birth hour's blot:
For marks descried in men's nativity
Are nature's faults, not their own infamy.
Lucrece.

A few quotations on abortion, and some others that are intimately related to obstetrics, remain.

If ever he have child, abortive be it,
Prodigious, and untimely brought to light,
Whose ugly and unnatural aspect
May fright the hopeful mother at the view.
Richard III., Act I., Sc. II.

Why should I joy in any abortive birth?
Love's Labour's Lost, Act I., Sc. I.

Truth is truth: large length of seas and shores
Between my father and my mother lay,—
And I have heard my father speak * * *
That this, my mother's son, was none of his;
And, if he were, he came into the world
Full fourteen weeks before the course of time.
King John, Act I., Sc. I.

Shakespeare has interwoven some of his family history here, and made the advent of Philip, the Bastard, correspond exactly to the untimely birth of his eldest daughter Susanna, who appeared only five and a half months after his marriage—"full fourteen weeks before the course of time." Later on in the play we find the following:

Your brother is legitimate,
Your father's wife did after wedlock bear him.

—thus furnishing proof of legitimacy in such cases.

She is, something before her time, deliver'd.
* * * A daughter; and a goodly babe,
Lusty, and like to live.
Winter's Tale, Act II., Sc. II.

O pray God, the fruit of her womb miscarry.
Henry IV—2d, Act. V., Sc. IV.

She had also snatch'd a moment since her marriage
To bear a son and heir —and one miscarriage.
                    Byron—Don Juan, Canto XIV., Verse LVI.

Macduff was from his mother's womb
Untimely ripp'd.
                    Macbeth, Act V., Sc. VIII.

Some griefs are med'cinable; that is, one of them,
For it doth physic love.
                    Cymbeline, Act III., Sc. II.

This bastard graff shall never come to growth :
He shall not boast who did thy stock pollute
That thou art doting father of his fruit.
                                        Lucrece.

Grant, that our hopes, (yet likely of fair birth)
Should be still-born. * * * *
                    Henry IV—2d, Act I., Sc. III.

The barren, touched in this holy chase,
Shake off their sterile curse.
                    Julius Cæsar, Act I., Sc. II.

This supposed charm against sterility, says Dyer, "is copied from Plutarch, who, in his description of the festival Lupercalia, tells us how 'noble young men run naked through the city, striking in sport whom they meet in the way with leather thongs,' which blows were commonly believed to have the wonderful effect attributed to them by Cæsar."

I had then laid wormwood to my dug,
* * * it did taste the wormwood on the nipple
Of my dug, and felt it bitter.
                    Romeo and Juliet, Act I., Sc. III.

I have given suck, and know
How tender 'tis to love the babe that milks me;
I would, while it was smiling in my face,
Have pluck'd my nipple from his boneless gums,
And dash'd the brains out, had I so sworn
As you have done to this.   *Macbeth, Act I., Sc. VII.*

Eggs, oysters too, are amatory food.
                    Byron—Don Juan, Canto, II., Verse CLXX.

Surely Byron knew of the stimulating qualities of eggs and oysters, and no doubt took them with as much faith as the worn-out debauchee of to-day does, as he sits down to his "plate of raw" and his "sherry and egg."

PART V.

PHYSIOLOGY.

Mr. Hackett, noticing the numerous allusions in Shakespeare to the blood, and to a circulation of this fluid to and from the heart or the liver, was led, in 1859, to express the absurd idea that William Shakespeare had anticipated Harvey in the discovery of the circulation of the blood.

" What damned error, but some sober brow
Will bless it, and approve it with a text."

Mr. Hackett found many thoughts in Shakespeare concerning the circulation which were applicable to Harvey's theory.

See, how the blood is settled in his face !
Oft have I seen a timely-parted ghost,
Of ashy semblance, meagre, pale and bloodless,
Being all descended to the labouring heart ;
Who, in the conflict that it holds with death,
Attracts the same for aidance 'gainst the enemy ;
Which with the heart there cools, and ne'er returneth
To blush and beautify the cheek again.
Henry VI—2d., Act III., Sc. II.

You are * * * *
As dear to me as are the ruddy drops
That visit my sad heart.
Julius Cæsar, Act II., Sc. I.

Why does my blood thus muster to my heart,
Making both it unable for itself,
And dispossessing all my other parts
Of necessary fitness?
Measure for Measure, Act II., Sc. IV.

My heart drops blood.
Cymbeline, Act V., Sc V.

I am sure my heart wept blood.
Winter's Tale, Act V., Sc. II.

73

These words of yours draw life-blood from my heart.

Henry VI., Act IV., Sc. VI.

The blood weeps from my heart.

Henry IV—2d, Act IV., Sc. IV.

I send it through the rivers of your blood,
Even to the court, the heart—to the seat o' the brain;
And, through the cranks and offices of man,
The strongest nerves and small inferior veins,
From me receive that natural competency
Whereby they live.

Coriolanus, Act I., Sc. I.

The tide of blood in me
Hath proudly flow'd in vanity, till now;
Now doth it turn, and ebb back to the sea,
Where it shall mingle with the state of floods,
And flow henceforth in formal majesty.

Henry IV—2d, Act V., Sc. II.

The spring, the head, the fountain of your blood
Is stopp'd; the very source of it is stopped.

Macbeth, Act II., Sc. II.

—— my heart, * * *
The fountain from the which my current runs,
Or else dries up.

Othello, Act IV., Sc. II.

I cannot rest
Until the white rose that I wear, be dy'd
Even in the lukewarm blood of Henry's heart.

Henry VI—3d, Act I., Sc. II.

Snakes, in my heart-blood warm'd, that sting my heart!

Richard II., Act III. Sc. II.

Thy heart-blood I will have for this day's work.

Henry VI., Act I., Sc. III.

Thou wouldst have left thy dearest heart-blood there,
Rather than have made that savage duke thine heir.

Henry VI—3d, Act I., Sc. I.

Her blue blood changed to black in every vein,
Wanting the spring that those shrunk pipes had fed,
Show'd life imprison'd in a body dead.

Lucrece.

Corrupted blood some watery token shows;
And blood untainted still doth red abide,
Blushing at that which is so putrefied.

Lucrece.

Even here she sheathed in her harmless breast
A harmful knife, * * * * * *
And bubbling from her breast, it doth divide
In two slow rivers, that the crimson blood
Circles her body in on every side, * * *
Some of her blood still pure and red remain'd,
And some look'd black.

Lucrece.

But are you flesh and blood?
Have you a working pulse?

Pericles, Act V., Sc. I.

I drink the air before me, and return
Or e'er your pulse twice beat.

Tempest, Act V., Sc. I.

My pulse as yours doth temperately keep time,
And makes as healthful music.

Hamlet, Act III., Sc. IV.

Your pulsidge beats as extraordinarily as heart would desire.

Henry IV—2d, Act II., Sc. IV.

Even as my life, or blood that fosters it.

Pericles, Act II., Sc. V.

Swift as quicksilver it courses through
The natural gates and alleys of the body.

Hamlet, Act I., Sc. V.

Shakespeare died in 1616. Harvey first published his theory
in 1619. It must be remembered that at this time many ideas
were afloat concerning the circulation. Among the older theories
were those of Hippocrates, Praxágoras, and Erasistratus, who
hold that the arteries contained air, and that, therefore, the *veins*
were the *only* blood-holding vessels, and that they had their
origin in the liver Galen, the most celebrated of ancient medical writers, who lived as early as 150 A. D. taught that the left
ventricle of the heart was the common origin of all arteries, and
that the arteries of living animals contained blood, *not* air; but
he did not advance with his studies so as to learn in what direc-
tion the blood flowed, or whether it was movable or stationary.

The distinguished Michael Servetus, who was burned with his books, by order of Calvin, in 1553, taught that the blood flowed from the right ventricle, through the pulmonary artery to the lungs, and thence through the plumonary vein and left auricle into the corresponding ventricle from which it was conveyed by the aorta to all parts of the body. Dr. Bucknill is of the opinion that Shakespeare followed Hippocrates in his theory that the veins were the only blood vessels and that they came from the liver. It is very evident, from the many allusions given below, that he did at different periods adhere to this belief.

> Let my liver rather heat with wine,
> Than my heart cool with mortifying groans.
> > *Merchant of Venice, Act I., Sc. I.*

For Andrew, if he were opened, and you find so much blood in his liver as will clog the foot of a flea, I'll eat the rest of the anatomy.
> > *Twelfth Night, Act III., Sc. II.*

> I'll empty all these veins,
> And shed my dear blood drop by drop.
> > *Henry IV., Act I., Sc. III.*

> I'll have more lives
> Than drops of blood were in my father's veins.
> > *Henry VI—3d, Act I., Sc. I.*

> Let me have
> A dram of poison ; such soon-speeding gear
> As will disperse itself through all the veins.
> > *Romeo and Juliet, Act V., Sc. I.*

> I freely told you, all the wealth I had
> Ran in my veins.
> > *Merchant of Venice, Act III., Sc. II.*

> The blood and courage that renowned them,
> Runs in your veins.
> > *Henry V., Act I., Sc. II.*

> —through all thy veins shall run
> A cold and drowsy humour, which shall seize
> Each vital spirit ; for no pulse shall keep
> His natural progress but surcease to beat.
> > *Romeo and Juliet, Act IV., Sc. I.*

> There is * * * * *
> Scarce blood enough in all their sickly veins.
> > *Henry V., Act IV., Sc. II.*

My blood speaks to you in my veins.
>> *Merchant of Venice, Act III., Sc. II.*

While warm life plays in that infant's veins.
>> *King John, Act III., Sc. IV.*

Had bak'd thy blood, and made it heavy thick,
Which, else, runs tickling up and down the veins.
>> *King John, Act III., Sc. III.*

'Tis thy presence that exhales this blood
From cold and empty veins, where no blood dwells.
>> *Richard III., Act I., Sc. II.*

Stuff'd within with bloody veins.
>> *Pericles, Act I., Sc. IV.*

For every false drop in her bawdy veins
A Grecian's life hath sunk.
>> *Troilus and Cressida, Act IV., Sc. I.*

If so thou yield him, there is gold, and here
My bluest veins to kiss.
>> *Antony and Cleopatra, Act II., Sc. V.*

That those veins
Did verily bear blood.
>> *Winter's Tale, Act V., Sc. III.*

The veins unfill'd, our blood is cold.
>> *Coriolanus, Act V., Sc. I.*

I have a faint cold, fear thrills through my veins
That almost freezes up the heat of life.
>> *Romeo and Juliet, Act IV., Sc. III.*

——purple fountains issuing from your veins.
>> *Romeo and Juliet, Act I., Sc. I.*

The arteries or "air pipes" were supposed, according to this theory of Hippocrates, to contain an ærial fluid.

These pipes and these conveyances of our blood.
>> *Coriolanus, Act V., Sc. I.*

Universal plodding poisons up
The nimble spirits in the arteries.
>> *Love's Labour's Lost, Act IV., Sc. III.*

My fate cries out,
And makes each petty artery in this body
As hardy as the Nemean lion's nerve.
>> *Hamlet, Act I., Sc. IV.*

It is more reasonable to suppose that Shakespeare did not tie himself down to any one theory concerning the circulation, but that sometimes he had in mind the theory of Michael Servetus, (to which all the heart allusions will apply), and at other times that of Hippocrates, (which accounts for all the thoughts regarding the liver as the propeller of the blood through the veins). The immortal Harvey was the first to point out the true idea of the circulation: the idea that the blood was forced by the heart through the arteries, a pure live-supporting fluid; that it went to the extreme parts of the body, giving nutriment, taking up impurities, and then returning by way of the veins to the heart,—thence to the lungs to be purified before being again sent out on it's life-sustaining journey. None of the quotations from Shakespeare express this idea, excepting perhaps one, and that rather vaguely.

> The tide of blood in me
> Hath proudly flow'd in vanity, till now;
> Now doth it turn, and ebb back to the sea,
> Where it shall mingle with the state of floods,
> And flow henceforth in formal majesty.
>
> *Henry IV—2d, Act V., Sc. II.*

We can not believe, however that he possessed the knowledge of Harvey's theory, and can only say in his own words:

> There is no vice so simple, but assumes
> Some mark of virtue on it's outward parts.

The physiology of the digestive system is excellently described in Coriolanus.

> *Men.* There was a time, when all the body's members
> Rebell'd against the belly; thus accus'd it:
> That only like a gulf it did remain
> I' the midst o' the body, idle and unactive,
> Still cupboarding the viand, never bearing
> Like labour with the rest, where the other instruments
> Did see, and hear, devise, instruct, walk, feel,
> And mutually participate, did minister
> Unto the appetite and affection common
> Of the whole body. The belly answer'd
> * * * * with a kind of smile,
> Which ne'er came from the lungs, but even thus,

For, look you, I may make the belly smile,
As well as speak,—it tauntingly replied
To the discontented members, the mutinous parts
That envied his receipt. * * *
* * * * * * * * *

1st Cit. Your belly's answer? What!
The kingly-crown'd head, the vigilant eye,
The counsellor heart, the arm our soldier,
Our steed the leg, the tongue our trumpeter,
With other muniments and petty helps
In this our fabric, if that they * * *
Should, by the cormorant belly be restrain'd,
Who is the sink o' the body.

Men. * * * * * * *
True it is, quoth the belly,
That I receive the general food at first,
Which you do live upon; and fit it is,
Because I am the store house and the shop
Of the whole body: but if you do remember,
I send it through the rivers of your blood,
Even to the court, the heart—to the seat o' the brain;
And, through the cranks and offices of man,
The strongest nerves and small inferior veins,
From me receive that natural competency
Whereby they live. *Act I., Sc. I.*

For your digestion's sake
An after-dinner speech.
Troilus and Cressida, Act II., Sc. III.

To make our appetites more keen,
With eager compounds we our palate urge.
Sonnets, CXVIII.

My cheese, my digestion.
Troilus and Cressida, Act II., Sc. III.

I say, whatever you maintain
Of Alma in the heart or brain,
The plainest man alive may tell ye
Her seat of empire is the belly.
From hence she sends out those supplies
Which make us either stout or wise;
Your stomach makes the fabric roll
Just as the bias rules the bowl.
The great Achilles might employ
The strength designed to ruin Troy;
He dined on lion's marrow, spread
On toast of ammunition bread;
But by his mother sent away

Amongst the Thracian girls to play,
Effeminate he sat and quiet—
Strange product of a cheese-cake diet!
Was ever Tartar fierce or cruel
Upon the strength of water-gruel?
But who shall stand his rage or force
If first he rides, then eats his horse?
Salads and eggs, and lighter fare,
Tunes the Italian spark's guitar;
And, if I take Dan Congrieve right,
Pudding and beef make Britons fight.
Tokay and coffee cause this work
Between the German and the Turk:
And both, as they provisions want,
Chicane, avoid, retire, and faint.
    *   *   *   *   *   *
But, spoil the organ of digestion,
And you entirely change the question:
Alma's affairs no power can mend;
The jest, alas! is at an end. * * *
                                                    Prior.

A few remaining physiological thoughts are interesting. As is well known, we are much better able to judge the size and distance of objects on the same level with us than we are when they are either above or below us. When we view objects from a height they appear much less than they would were we at the same distance from them on the same level. Shakespeare has beautifully shown this effect in King Lear.

How fearful
And dizzy 'tis, to cast one's eyes so low!
The crows, and choughs, that wing the midway air,
Show scarce so gross as beetles. Half way down
Hangs one that gathers samphire; dreadful trade!
Methinks, he seems no bigger than his head:
The fishermen that walk upon the beach,
Appear like mice. * * * *   *Act IV., Sc. VI.*

The subject of pupillary reflexes has received mention by many of the older writers. It was a source of amusement to lovers in the old time to look into each others eyes in search of their own reflection.

Joy had the like conception in our eyes,
And, at that instant, like a babe, sprung up.
                              Timon of Athens, Act I., Sc. II.

Look in my eyes, my blushing fair,
Thou'lt see thyself reflected there;
As I gaze on thine, I see
Two little miniatures of me.

Thus in our looks some propagation lies,
For we make babies in each other's eyes.   *Tom Moore.*

When a young lady wrings you by the hand, thus,
Or with an amorous touch presses your foot ;
Looks babies in your eyes, plays with your locks.
*   *   *   *   *   *   *   *   *
                Massinger—Renegado, Act II., Sc, IV.

It has been a view long held that the height of the forehead
is an index of the intellectual character of the individual.
Shakespeare has referred to this in several plays.

We shall lose our time,
And all be turn'd to barnacles, or to apes,
With foreheads villainous low.   *Tempest, Act IV., Sc. I.*

Ay, but her forehead's low, as mine's as high.
         Two Gentlemen of Verona, Act IV., Sc IV.

Cleopatra. Bear'st thou her face in mind ? is't long or round ?
Messenger. Round, even to faultiness.
Cleopatra. For the most part too,
     They are foolish that are so. Her hair, what colour ?
Messenger. Brown, madame, and her forehead
     As low as you would wish it.
         Antony and Cleopatra, Act III., Sc. III.

The old superstition that much hair on the head indicated a
want of intellect is alluded to in Two Gentlemen of Verona.

Speed. Item, *she hath more hair than wit.*
Laun. More hair than wit,—it may be ; I'll prove it: the cover of the salt
     hides the salt, and therefore it is more than the salt; the hair that
     covers the wit is more than the wit; for the greater hides the less.
         Act III., Sc. I.

Ant. S. Why is Time such a niggard of hair, being, as it is, so plentiful an
     excrement ?
Dro. S. Because it is a blessing that he bestows on beasts ; and what he hath
     scanted men in hair he hath given them in wit.
Ant. S. Why, but there's many a man hath more hair than wit.
Dro. S. Not a man of those but he hath the wit to lose his hair.
Ant. S. Why, thou did'st conclude hairy men plain dealers without wit.
         Comedy of Errors, Act II., Sc. II.

This great voluminous pamphlet may be said
To be like one that hath more hair than head ;
More excrement than body : trees which sprout
With broadest leaves have still the smallest fruit.   *Suckling—Aglaura.*

He had some idea of the sympathetic connection between the
organs of the body, and has furnished us with a good example

81

of superstition connected with sympathy. It was an old superstition that the wounds of a murdered person would bleed afresh
if the body was touched by the murderer, and this has nicely
been brought out in Richard III.

> O, gentlemen, see, see! dead Henry's wounds
> Open their congeal'd mouths and bleed afresh !
> Blush, blush, thou lump of foul deformity ;
> For 'tis thy presence that exhales this blood
> From cold and empty veins, where no blood dwells,
> Thy deed, inhuman and unnatural,
> Provokes this deluge most unnatural.   *Act I., Sc. II.*

Dunglison explains these superstitions " either on purely physical principles, or on the excited imagination of the observer,"
and cites two interesting cases—one attested by John Demarest,
coroner of Bergen county, New Jersey, (1767), and the other
which occurred near Easton, Pennsylvania. Of the latter case
he says : " The superstition has, indeed, its believers among us.
On the trial of Getter, who was executed about five years ago
(1833) in Pennsylvania, for the murder of his wife, a female wit-
ness deposed on oath as follows : ' If my throat was to be cut,
I could tell, before God Almighty, that the deceased smiled when
he (the murderer) touched her. I swore this before the justices,
and that she bled considerably. I was sent for to dress her and
lay her out. He touched her twice. He made no hesitation
about doing it. I also swore before the justice that it was
observed by other people in the house.' " Dyer cites a number
of similar cases, and quotes the following as a supposed cause of
the phenomenon from the "Athenian Oracle," (1–106) : " The
blood is congealed in the body for two or three days, and then
becomes liquid again, in its tendency to corruption. The air
being heated by many persons coming about the body is the
same thing to it as motion is. 'Tis observed that dead bodies
will bleed in a concourse of people, when murderers are absent
as well as present, yet legislators have thought it fit to authorize
it, and use this trial as an argument, at least to frighten, though
'tis no conclusive one to condemn them." The practice, however, caused many an innocent spectator to receive the fatal
penalty.

PART VI.

ANATOMY.

Anatomy received some attention.

Ant. S. What's her name?

Dro. S. Nell, sir; but her name and three quarters, that's an ell and three-quarters, will not measure her from hip to hip.

Ant. S. Then she bears some breadth?

Dro. S No longer from head to foot than from hip to hip; she is spherical like a globe,—I could find out countries on her.

Ant. S. In what part of her body stands Ireland?

Dro. S. Marry, sir, in her buttocks; I found it out by the bogs.

Ant. S. Where's Scotland?

Dro. S. I found it by the barrenness; hard, in the palm of the hand.

Ant. S. Where's France?

Dro. S. In her forehead; arm'd and reverted, making war against her heir.

Ant. S. Where's England?

Dro. S. I looked for the chalky cliffs, but I could find no whiteness in them; but I guess it stood in her chin, by the salt rheum that ran between France and it.

Ant. S. Where's Spain?

Dro. S. Faith, I saw it not; but I felt it hot in her breath.

Ant. S. Where's America, the Indies?

Dro. S. O, sir, upon her nose,—all o'er embellished with rubies, carbuncles, saphires, declining their rich aspect to the hot breath of Spain, who sent whole armadoes of carracks to be ballast at her nose.

Ant. S. Where stood Belgia, the Netherlands?

Dro. S. O, sir, I did not look so low. * * *

Comedy of Errors. Act III., Sc. II.

Feed where thou wilt, on mountain or in dale:
Graze on my lips; and if those hills be dry,
Stray lower, where the pleasant fountains lie.
Within this limit is relief enough,
Sweet bottom-grass, and high delightful plain,
Round rising hillocks, brakes obscure and rough,
To shelter thee from tempest and from rain:
Then be my deer, since I am such a park;
No dog shall rouse thee, though a thousand bark.

Venus and Adonis.

The old superstition that our bodies consisted of the elements—fire, water, earth and air—has been mentioned.

Sir Toby. Does not our life consist of four elements?
Sir Andrew. 'Faith so they say; but I think it rather consists of eating and
drinking. *Twelfth Night, Act II., Sc. III.*

His life was gentle; and the elements
So mix'd in him, that nature might stand up,
And say to all the world, *This was a man!*
Julius Cæsar, Act V., Sc. V.

I am fire and air; my other elements
I give to baser life. *Antony and Cleopatra, Act V., Sc. II.*

O tell me, friar, tell me,
In what vile part of this anatomy
Doth my name lodge? *Romeo and Juliet, Act III., Sc. III.*

The brain was thought only to have three ventricles by the
old anatomists; what is now the fourth ventricle was called by
them the third, and was supposed to be the seat of memory.

A foolish extravagant spirit, full of forms, figures, shapes, objects, ideas,
apprehensions, motions, revolutions: these are begot in the ventricle of memory, nourished in the womb of *pia mater.*
Love's Labour's Lost, Act IV., Sc. II.

—whose skull Jove cram with brains!
* * * * has a most weak *pia mater.*
Twelfth Night, Act I., Sc. V.

Many a time, but for a sallet, my brain-pan had been cleft with a brown bill.
Henry VI—2d, Act IV., Sc. X.

Servant. My lord you have one eye left.
Cornwall. Lest it see more, prevent it.—
Out, vile jelly!
Where is thy lustre now?
King Lear, Act III., Sc. VII.

Like a strutting player,—whose conceit
Lies in his hamstring. *Troilus and Cressida, Act I., Sc. III.*

Thy bones are hollow.
Measure for Measure, Act I., Sc. II.

Thy bones are marrowless. *Macbeth, Act III., Sc. IV.*

A dying Moslem, who had felt the foot
Of a foe o'er him, snatch'd at it, and bit
The very tendon which is most acute—
(That which some ancient muse or modern wit
Named after thee Achilles) and quite through't
He made the teeth meet.
Byron—Don Juan, Canto VIII., Verse LXXXIV.

PART VII.

Pharmacy was not overlooked.

I do remember an apothecary,—
And hereabouts he dwells,—which late I noted
In tatter'd weeds, with overwhelming brows,
Culling of simples: meagre were his looks,
Sharp misery had worn him to the bones;
And in his needy shop a tortoise hung,
An alligator stuff'd, and other skins
Of ill-shap'd fishes; and, about his shelves,
A beggarly account of empty boxes,
Green earthen pots, bladders, and musty seeds,
Remnants of packthread, and old cakes of roses,
Were thinly scatter'd to make up a show.
Noting this penury, to myself I said—
An if a man need poison now,
Whose sale is present death in Mantua,
Here lives a caitiff wretch would sell it him.
 *   *   *   *   *   *   *   *
What, ho! apothecary!

Romeo and Juliet, Act V., Sc. I.

O, true apothecary!
Thy drugs are quick.

Romeo and Juliet, Act V., Sc. III.

He did buy a poison of a poor apothecary,
And there withal came to this vault to die.

Romeo and Juliet, Act V., Sc. III.

Bid the apothecary
Bring the strong poison that I bought of him.

Henry VI—2d, Act III., Sc. III.

Your master will be dead ere you return;
There's nothing can be minister'd to nature.
That can recover him. Give this to the 'pothecary,
And tell me how it works.

Pericles, Act III., Sc. II.

Great griefs, I see, medicine the less.

Cymbeline, Act IV., Sc. II.

85

That drug-damn'd Italy hath out-craftied him.
Cymbeline, Act III., Sc. IV.

One, whose subdu'd eyes,
Albeit unused to the melting mood,
Drop tears as fast as the Arabian trees
Their med'cinable gum.
Othello, Act V., Sc. II.

Set ratsbane by his porridge.
King Lear, Act III., Sc. IV.

I had as lief they would put ratsbane in my mouth, as offer to stop it with security.
Henry IV—2d, Act I., Sc. II.

I would the milk
Thy mother gave thee, when thou suck'dst her breast,
Had been a little ratsbane for thy sake!
Henry VI., Act V., Sc. IV.

If you have poison for me I will drink it.
King Lear, Act IV., Sc. VII.

I have bought the oil, the balsamum and aqua-vitæ.
Comedy of Errors, Act IV., Sc. I.

Give me some aqua-vitæ.
Romeo and Juliet, Act III., Sc. II. ✕

www.ingramcontent.com/pod-product-compliance
Lightning Source LLC
Chambersburg PA
CBHW020304090426
42735CB00009B/1219